Lecture Notes in Mathematics

Edited by J.-M. Morel, F. Takens and B. Teissier

Editorial Policy
for the publication of monographs

1. Lecture Notes aim to report new developments in all areas of mathematics – quickly, informally and at a high level. Monograph manuscripts should be reasonably self-contained and rounded off. Thus they may, and often will, present not only results of the author but also related work by other people. They may be based on specialized lecture courses. Furthermore, the manuscripts should provide sufficient motivation, examples and applications. This clearly distinguishes Lecture Notes from journal articles or technical reports which normally are very concise. Articles intended for a journal but too long to be accepted by most journals, usually do not have this "lecture notes" character. For similar reasons it is unusual for doctoral theses to be accepted for the Lecture Notes series.

2. Manuscripts should be submitted (preferably in duplicate) either to one of the series editors or to Springer-Verlag, Heidelberg. In general, manuscripts will be sent out to 2 external referees for evaluation. If a decision cannot yet be reached on the basis of the first 2 reports, further referees may be contacted: the author will be informed of this. A final decision to publish can be made only on the basis of the complete manuscript, however a refereeing process leading to a preliminary decision can be based on a pre-final or incomplete manuscript. The strict minimum amount of material that will be considered should include a detailed outline describing the planned contents of each chapter, a bibliography and several sample chapters.
Authors should be aware that incomplete or insufficiently close to final manuscripts almost always result in longer refereeing times and nevertheless unclear referees' recommendations, making further refereeing of a final draft necessary.
Authors should also be aware that parallel submission of their manuscript to another publisher while under consideration for LNM will in general lead to immediate rejection.

3. Manuscripts should in general be submitted in English.
Final manuscripts should contain at least 100 pages of mathematical text and should include
– a table of contents;
– an informative introduction, with adequate motivation and perhaps some historical remarks: it should be accessible to a reader not intimately familiar with the topic treated;
– a subject index: as a rule this is genuinely helpful for the reader.

Continued on inside back-cover

Lecture Notes in Mathematics

1777

Editors:
J.-M. Morel, Cachan
F. Takens, Groningen
B. Teissier, Paris

Springer
Berlin
Heidelberg
New York
Barcelona
Hong Kong
London
Milan
Paris
Tokyo

Eduardo García-Río
Demir N. Kupeli
Ramón Vázquez-Lorenzo

Osserman Manifolds in Semi-Riemannian Geometry

 Springer

Authors

Eduardo García-Río

Dept. of Geometry and Topology
Faculty of Mathematics
Univ. of Santiago de Compostela
15782 Santiago, Spain
E-mail: xtedugr@usc.es

Demir N. Kupeli

Dept. of Mathematics
Atilim University
Incek
06836 Ankara, Turkey
E-mail: dnkupeli@superonline.com

Ramón Vázquez-Lorenzo

Dept. of Geometry and Topology
Faculty of Mathematics
Univ. of Santiago de Compostela
15782 Santiago, Spain
E-mail: ravazlor@edu.xunta.es

Cataloging-in-Publication Data applied for.

Die Deutsche Bibliothek - CIP-Einheitsaufnahme

García-Río, Eduardo:
Osserman manifolds in semi-Riemannian geometry / Eduardo García-Río ; Demir
N. Kupeli ; Ramón Vázquez-Lorenzo. - Berlin ; Heidelberg ; New York ;
Barcelona ; Hong Kong ; London ; Milan ; Paris ; Tokyo : Springer, 2002
  (Lecture notes in mathematics ; 1777)
  ISBN 3-540-43144-6

Mathematics Subject Classification (2000):
53C25, 53C50, 53C55, 53C80

ISSN 0075-8434
ISBN 3-540-43144-6 Springer-Verlag Berlin Heidelberg New York

Springer-Verlag Berlin Heidelberg New York a member of BertelsmannSpringer
Science + Business Media GmbH

http://www.springer.de

© Springer-Verlag Berlin Heidelberg 2002
Printed in Germany

The use of general descriptive names, registered names, trademarks, etc. in this publication does not imply,
even in the absence of a specific statement, that such names are exempt from the relevant protective laws
and regulations and therefore free for general use.

Typesetting: Camera-ready TEX output by the authors

SPIN: 10864804   41/3142-543210 - Printed on acid-free paper

To Fernanda, Meltem and Maica

Preface

The notion of curvature is one of the central concepts of differential geometry, perhaps the central one, which distinguishes the geometric core of the subject from the others that are analytic, algebraic or topological. It has always been a pursuit of great interest to understand to what extent the sectional curvatures of a semi-Riemannian manifold can provide information about the curvature and metric tensors. One question of this kind that has been under much scrutiny in recent years, which will be our central theme, is whether the Osserman condition (involving sectional curvatures, and more precisely, the Jacobi operator) determines the curvature and metric tensors. In this research monograph, our goal is to expound the recent developments in the exploration of the answer to this question in Riemannian, Lorentzian, semi-Riemannian and affine differential geometry.

The question in its original form is known as the Osserman conjecture in Riemannian geometry. Significant progress has been made in search of a complete answer to this question by Chi [39], [40], [41], albeit it still remains largely open. Later the conjecture was questioned in Lorentzian geometry in its Lorentzian setting. An affirmative answer to the Osserman conjecture was obtained in a sequence of papers by García-Río and Kupeli [58], García-Río, Kupeli and Vázquez-Abal [59] and Blažić, Bokan and Gilkey [16]. Finally the conjecture appealed in semi-Riemannian geometry in its semi-Riemannian forms. This time, however, several counterexamples were found to the conjecture in [23], [31], [61], which diverted the research to the direction of understanding the semi-Riemannian Osserman manifolds. On the other hand, Osserman condition can be stated in affine differential geometry naturally as well and the affine Osserman manifolds were lately investigated in [60].

We plan to present all these developments in six chapters. Preliminaries needed in this monograph are intermediate level aspects of differential geometry. Instead of collecting them in a single chapter, we prefer to introduce them briefly as they are encountered in each section. We expect that this approach will lead the reader quickly into the subject and enable them to read the chapters independently without having to know all the prerequisites before hand.

In Chapter 1, we give the main definitions of this monograph in their general setting in semi-Riemannian geometry. Corresponding definitions for the Riemannian and Lorentzian counterparts are special cases of them. We then prove that the timelike and spacelike Osserman conditions are equivalent, with respect to which we define a semi-Riemannian manifold to be Osserman if the characteristic polynomials of the Jacobi operators are independent of the vectors in both the timelike and spacelike unit tangent bundles of the manifold. We illustrate some examples that serve as model spaces of semi-Riemannian Osserman manifolds and prove a result to relate the pointwise and (global) Osserman conditions, where the pointwise Osserman condition is the version of the global definition when the "tangent bundle" is replaced by "tangent space at each point".

In Chapter 2, we study the known results toward an affirmative answer to the Osserman conjecture in Riemannian geometry, which states that a Riemannian Osserman manifold is either flat or rank-one symmetric. We demonstrate the solution to this conjecture in full for the dimensions $n \neq 4m$, $m \geq 1$, and outline the solutions for the other dimensions under certain additional assumptions.

In Chapter 3, we provide the complete solution to Osserman conjecture in Lorentzian geometry. That is, we show that a Lorentzian Osserman manifold is a real space form, and hence is either flat or rank-one symmetric. The proof of this fact is different from what has appeared in the literature in that we explore the equivalence between the timelike and spacelike Osserman conditions to make the proof rather simple and short. We also introduce the null Osserman condition and statements equivalent to it, and show that it yields certain warped product decomposition theorems under additional assumptions.

In Chapter 4, we study semi-Riemannian Osserman manifolds of signature $(2, 2)$. First we construct counterexamples to the Osserman conjecture in semi-Riemannian geometry by showing the existence of nonsymmetric semi-Riemannian Osserman manifolds of signature $(2, 2)$, some of which are not even locally homogeneous. The second part of this chapter is devoted to a study of those semi-Riemannian manifolds satisfying the Osserman condition by following the work of Blažić, Bokan and Rakić [18]. Here the Osserman curvature tensors are classified into four types according to the properties of the minimal and characteristic polynomials of the Jacobi operators. (In fact, this is a generalization to the semi-Riemannian setting of the equivalence between pointwise Osserman and Einstein self-dual 4-manifolds previously pointed out in the Riemannian case.) Based on this information we obtain a classification of semi-Riemannian manifolds with metric tensors of signature $(2, 2)$ whose Jacobi operators are diagonalizable. They must be either flat or locally rank-one symmetric, which is contrary to the nondiagonalizable case where two kinds of rank-two symmetric spaces are allowed.

Included in Chapter 5 are new classification results of ours for higher-dimensional semi-Riemannian Osserman manifolds. Due to the existence of non-symmetric examples in any signature (p, q), $p, q > 1$, we focus on the examination of the simplest cases of semi-Riemannian Osserman manifolds. Following the work of Chi [41] in the Riemannian case, we consider semi-Riemannian Osserman manifolds with exactly two distinct eigenvalues whose associated eigenspaces satisfy a kind of infinitesimal Hopf fibration property and call them semi-Riemannian special Osserman manifolds. The main objective of this chapter is to prove that such manifolds are either locally complex, quaternionic, paracomplex or paraquaternionic space forms, or locally isometric to the Cayley planes over the octonians or the anti-octonians. This shows that, besides the space forms, semi-Riemannian special Osserman manifolds correspond to the simplest semi-Riemannian manifolds from the viewpoint of their curvatures.

In Chapter 6, we review certain Osserman-related conditions. Following [133], [67] and [76] we introduce semi-Riemannian generalized Osserman manifolds. The fact that the space forms are the only semi-Riemannian generalized Osserman rank-one symmetric spaces shows that such a condition is much more restrictive than the Osserman condition. Along a different vein, the Osserman condition finds its natural setting in affine differential geometry as well. The notion of the affine Osserman condition is originated from our effort to supply new examples of semi-Riemannian Osserman manifolds via the construction called the Riemann extension [60]. It turns out that the affine Osserman condition does not generalize the semi-Riemannian ones since all eigenvalues of the Jacobi operators can be shown to be zero. It seems that the affine Osserman condition is the only reasonable notion of Osserman type in the general affine case where the unit sphere cannot be defined. Also, we relate ℭ-spaces and the Riemannian manifolds with isoparametric geodesic spheres to Riemannian Osserman manifolds here in connection with harmonic manifolds. To conclude, a kind of Osserman condition is studied for skew-symmetric curvature operators. Such a condition is shown to be the characteristic of space forms and certain special classes of Robertson-Walker type warped products in many cases. We only indicate here the basic notion and refer to [74] and [85] for the proofs and further references.

This work was supported through the grants DGESIC (TXT99-1482) and XUGA (PGIDT01PXIB) from Spain. The second-named author (DNK) is grateful to the Department of Geometry and Topology of the University of Santiago de Compostela for their kind invitation and for providing support while writing this book.

Santiago de Compostela, Ankara
May 2001

Eduardo García-Río, Demir N. Kupeli, Ramón Vázquez-Lorenzo

Contents

1. The Osserman Conditions in Semi-Riemannian Geometry

In this chapter we introduce the basic notation and terminology which are used throughout this book. In Section 1.1, we give the definition of Jacobi operator and its relation to curvatures. In Section 1.2, we define the timelike and spacelike Osserman conditions at a point and show their equivalence. Also by using this equivalence, we give the definition of Osserman condition at a point. In Section 1.3, we give the definitions of semi-Riemannian pointwise and global Osserman manifolds and a sufficient condition for a pointwise Osserman manifold to be globally Osserman is studied under some assumptions on the number of different eigenvalues of the Jacobi operators. In this section, we also give some model semi-Riemannian global Osserman manifolds.

Throughout this chapter, let (M, g) be a semi-Riemannian manifold of dimension $n \geq 2$ and index ν. That is, g is a metric tensor on M with signature (ν, η), where $\eta = n - \nu$.

1.1 The Jacobi Operator

First we fix some notation and terminology.

Definition 1.1.1. *Let (M, g) be a semi-Riemannian manifold. A nonzero vector $z \in T_p M$ is called:*

a) *timelike if $g(z, z) < 0$,*
b) *spacelike if $g(z, z) > 0$,*
c) *null if $g(z, z) = 0$,*
d) *nonnull if $g(z, z) \neq 0$.*

Also a nonnull vector z is called a unit vector if $| g(z, z) | = 1$.

Let $S_p^-(M)$, $S_p^+(M)$ and $S_p(M)$ be the sets of unit timelike, spacelike and nonnull vectors in $T_p M$, respectively. That is,

$$S_p^-(M) = \{z \in T_p M / g(z, z) = -1\},$$
$$S_p^+(M) = \{z \in T_p M / g(z, z) = 1\},$$
$$S_p(M) = \{z \in T_p M / \mid g(z, z) \mid = 1\} = S_p^-(M) \cup S_p^+(M).$$

Note that if (M, g) is not Riemannian (that is, $\nu \neq 0, n$) then $S_p(M)$ is not compact.

Definition 1.1.2. *Let (M, g) be a semi-Riemannian manifold. Then*

a) $S^-(M) = \bigcup_{p \in M} S_p^-(M) = \{z \in TM / g(z, z) = -1\}$ *is called the unit timelike bundle of (M, g).*

b) $S^+(M) = \bigcup_{p \in M} S_p^+(M) = \{z \in TM / g(z, z) = 1\}$ *is called the unit space-like bundle of (M, g).*

c) $S(M) = \bigcup_{p \in M} S_p(M) = \{z \in TM / \mid g(z, z) \mid = 1\}$ *is called the unit nonnull bundle of (M, g).*

Now we are ready to define the Jacobi operator.

Let (M, g) be a semi-Riemannian manifold and ∇ the Levi-Civita connection of (M, g). We define the *curvature tensor R* of ∇ by

$$R(X, Y)Z = \nabla_X \nabla_Y Z - \nabla_Y \nabla_X Z - \nabla_{[X,Y]} Z,$$

where X, Y, $Z \in \Gamma TM$ are vector fields on M and $[\,,\,]$ is the Lie bracket. Let $z \in T_p M$ and let

$$R(\,\cdot\,, z)z : T_p M \to T_p M$$

be the linear map defined by $(R(\,\cdot\,, z)z)x = R(x, z)z$. Note that, by curvature identities, since $g(R(x, z)z, z) = 0$, we have

$$R(\,\cdot\,, z)z : T_p M \to z^\perp,$$

where $z^\perp$ is the orthogonal space to $span\{z\}$. Now using the linear map $R(\,\cdot\,, z)z$, we define the Jacobi operator with respect to z. Note that, if $z \in S(M)$ then $z^\perp$ is a nondegenerate hyperspace in $T_p M$ (that is, the restriction of g to $z^\perp$ is nondegenerate), where $z \in S_p(M)$.

Definition 1.1.3. *Let (M, g) be a semi-Riemannian manifold and $z \in S(M)$. Then the restriction $R_z : z^\perp \to z^\perp$ of the linear map $R(\,\cdot\,, z)z$ to $z^\perp$ is called the Jacobi operator with respect to z, that is*

$$R_z x = R(x, z)z,$$

where $x \in z^\perp$.

Now se state some properties of the Jacobi operator.

Proposition 1.1.1. *Let (M, g) be a semi-Riemannian manifold and $z \in S(M)$. Then the Jacobi operator $R_z : z^\perp \to z^\perp$ is a self-adjoint map.*

Proof. Let $x, y \in z^{\perp}$. Then by curvature identities,

$$g(R_z x, y) = g(R(x, z)z, y) = g(R(z, y)x, z)$$
$$= g(R(y, z)z, x) = g(R_z y, x).$$

Hence R_z is self-adjoint. $\qquad\qquad\qquad\qquad\qquad\qquad\qquad\qquad\square$

Remark 1.1.1. Let (M, g) be a semi-Riemannian manifold, $z \in S(M)$ and R_z the Jacobi operator. Then note that,

$$trace R_z = \sum_{i=1}^{n-1} g(x_i, x_i) g(R_z x_i, x_i)$$

$$= \sum_{i=1}^{n-1} g(x_i, x_i) g(R(x_i, z)z, x_i)$$

$$= Ric(z, z),$$

where $\{x_1, \ldots, x_{n-1}\}$ is an orthonormal basis for $z^{\perp}$. Also for every nonnull unit $x \in z^{\perp}$, $P = span\{x, z\}$ is a nondegenerate plane in $T_p M$ (that is, the restriction of g to P is nondegenerate), where $z \in T_p M$, and the curvature $\kappa(P)$ of P is given by

$$\kappa(P) = \frac{g(R(x, z)z, x)}{g(x, x)g(z, z) - g(x, z)^2} = \frac{g(R_z x, x)}{g(x, x)g(z, z)}.$$

Hence, in Riemannian geometry, the eigenvalues of R_x represent the extremal values of the sectional curvatures of all planes containing x.

1.2 The Timelike and Spacelike Osserman Conditions at a Point

Let (M, g) be a semi-Riemannian manifold and $z \in S(M)$. Then the Jacobi operator $R_z : z^{\perp} \to z^{\perp}$ is a self-adjoint linear map. But in general, since $z^{\perp}$ has an induced indefinite inner product, R_z may not be diagonalizable. That is why we state Osserman conditions in terms of the characteristic polynomial of R_z rather than its eigenvalues as in Riemannian geometry. As we will remark later, both statements of Osserman conditions in terms of characteristic polynomials and eigenvalues of R_z coincide in Riemannian geometry.

Definition 1.2.1. *Let (M, g) be a semi-Riemannian manifold and $p \in M$.*

a) (M, g) is called timelike Osserman at p if the characteristic polynomial of R_z is independent of $z \in S_p^{-}(M)$.

b) (M, g) is called spacelike Osserman at p if the characteristic polynomial
of R_z is independent of $z \in S_p^+(M)$.

Remark 1.2.1. It is important to note that the Osserman condition could
be equivalently stated in terms of the constancy of the (possibly complex)
eigenvalues of the Jacobi operators, counted with multiplicities.

The fact that the Jacobi operators are, in general, nondiagonalizable in
the semi-Riemannian setting motivated the study of their normal forms (see
Section 4.2). Here it is worth to emphasize the role played by the minimal
polynomial of the Jacobi operators, since they may have nonconstant roots
even if the manifold is assumed to be Osserman as pointed out in the examples
in sections 4.1 and 5.1. Such a behaviour does not affect Riemannian nor
Lorentzian Osserman manifolds since their Jacobi operators are completely
determined by the knowledge of the corresponding eigenvalues.

Now we show that (M, g) being timelike Osserman at p is equivalent
to (M, g) being spacelike Osserman at p. Note in the proof, however, that
this equivalence does not imply that the characteristic polynomial of R_z is
independent of $z \in S(M)$.

Theorem 1.2.1. [60] *Let (M, g) be a semi-Riemannian manifold and $p \in M$. Then (M, g) is timelike Osserman at p if and only if (M, g) is spacelike Osserman at p.*

Proof. Let (M, g) be timelike Osserman at p and $f_z(t) = t^{n-1} + a_{n-2}t^{n-2} + \cdots + a_1 t + a_0$ be the characteristic polynomial of R_z for all $z \in S_p^-(M)$, where
$a_{n-2}, \ldots, a_1, a_0 \in \mathbb{R}$. Then the characteristic polynomial of $R(\cdot, z)z : T_pM \to z^\perp \subset T_pM$ is also independent of $z \in S^-(M)$. In fact, the characteristic
polynomial of $R(\cdot, z)z$ is then $h_z(t) = t^n + a_{n-2}t^{n-1} + \cdots + a_1 t^2 + a_0 t$ for
all $z \in S_p^-(M)$.

First note that, at $p \in M$, the metric tensor g and the curvature tensor R
are analytic functions on T_pM. Now complexify T_pM to $T_p^{\mathbb{C}}M$ and extend g
and R to be complex linear $g^{\mathbb{C}}$ and $R^{\mathbb{C}}$ on $T_p^{\mathbb{C}}M$, respectively. Next note that
$V = \{\gamma \in T_p^{\mathbb{C}}M / g^{\mathbb{C}}(\gamma, \gamma) \neq 0\}$ is a connected open dense subset of $T_p^{\mathbb{C}}M$ and
define $R_\gamma^{\mathbb{C}} : T_p^{\mathbb{C}}M \to T_p^{\mathbb{C}}M$ for each $\gamma \in V$ by

$$R_\gamma^{\mathbb{C}} \alpha = \frac{R^{\mathbb{C}}(\alpha, \gamma)\gamma}{g^{\mathbb{C}}(\gamma, \gamma)}.$$

Let $h_\gamma(t) = t^n + A_{n-2}(\gamma)t^{n-1} + \cdots + A_1(\gamma)t^2 + A_0(\gamma)t$ be the characteristic
polynomial of $R_\gamma^{\mathbb{C}}$ for each $\gamma \in V$, where $A_{n-2}, \ldots, A_1, A_0$ are functions on
$T_p^{\mathbb{C}}M$. Since $R^{\mathbb{C}}$ and $g^{\mathbb{C}}$ are analytic on $T_p^{\mathbb{C}}M$, clearly $A_{n-2}, \ldots, A_1, A_0$ are
also analytic on $T_p^{\mathbb{C}}M$.

Furthermore, let U^- and U^+ be the sets of timelike and spacelike vectors
in T_pM, respectively. We have $V \cap T_pM = U^- \cup U^+$.

Let $z \in S_p(M) \subset V$. Then note that $g^{\mathbb{C}}(z, z)R_z^{\mathbb{C}}$ is the complex linear extension of $R(\,\cdot\,, z)z$ to $T_p^{\mathbb{C}}M$ and hence, $g^{\mathbb{C}}(z, z)R_z^{\mathbb{C}}$ and $R(\,\cdot\,, z)z$ have the same characteristic polynomials. Now since (M, g) is timelike Osserman at p, it follows that the characteristic polynomial of $g^{\mathbb{C}}(z, z)R_z^{\mathbb{C}}$ is independent of $z \in S_p^-(M) \subset V$. Thus, since the coefficients of the characteristic polynomial of $g^{\mathbb{C}}(z, z)R_z^{\mathbb{C}}$ are analytic, they are constant for all $\gamma \in V$. Hence the coefficients $A_{n-2}, \ldots, A_1, A_0$ of the characteristic polynomial $R_\gamma^{\mathbb{C}}$ are constant for all $\gamma \in V$. This immediately implies that the coefficients of $g^{\mathbb{C}}(z, z)R_z^{\mathbb{C}}$ are independent of $z \in S_p^+(M)$ and hence, coefficients of the characteristic polynomial of $R(\,\cdot\,, z)z$ are independent of $z \in S_p^+(M)$. Thus the characteristic polynomial of R_z is independent of $z \in S_p^+(M)$. That is, (M, g) is spacelike Osserman at p. The converse is obtained in the same way. $\qquad\square$

Now by Theorem 1.2.1, the following definition can be given without loss of generality.

Definition 1.2.2. *Let (M, g) be a semi-Riemannian manifold. Then (M, g) is called Osserman at $p \in M$ if (M, g) is both timelike and spacelike Osserman at p.*

Remark 1.2.2. Note that the above definition *does not* mean that the characteristic polynomial of R_z is independent of $z \in S_p(M)$.

Next we show that, being Osserman for a semi-Riemannian manifold at a point simplifies the geometry at that point.

Lemma 1.2.1. [44] *Let $(V, \langle,\rangle)$ be an indefinite inner product space and let h be a bilinear form on V.*

 a) *If $h(u, u) = 0$ for every null $u \in V$ then $h = \lambda\langle,\rangle$, where $\lambda \in \mathbb{R}$.*
 b) *If $\mid h(x, x) \mid \leq d \in \mathbb{R}$ for every unit timelike vector $x \in V$ (or for every unit spacelike vector $x \in V$) then $h = \lambda\langle,\rangle$, where $\lambda \in \mathbb{R}$.*
 c) *If $h(x, x) \leq d_1 \in \mathbb{R}$ for every unit timelike (resp., spacelike) vector $x \in V$ and $h(y, y) \geq d_2 \in \mathbb{R}$ for every unit spacelike (resp., timelike) vector $y \in V$ then $h = \lambda\langle,\rangle$, where $\lambda \in \mathbb{R}$.*

Proof. See for example [44] and [101]. $\qquad\square$

Proposition 1.2.1. *Let (M, g) be a semi-Riemannian manifold. If (M, g) is Osserman at $p \in M$ then (M, g) is Einstein at $p \in M$, that is, $Ric = \lambda g$ at $p \in M$, where $\lambda \in \mathbb{R}$.*

Proof. Consider (M, g) as timelike Osserman at p. Then the characteristic polynomial $f_z(t) = \sum\limits_{i=0}^{n-1} a_i t^i$ of the Jacobi operator R_z is independent of $z \in S_p^-(M)$, where $a_{n-1} = 1$. In particular, since $trace R_z = -a_{n-2}$ and

$Ric(z, z) = traceR_z$, it follows that $Ric(z, z) = -a_{n-2}$ and hence is independent of $z \in S_p^-(M)$.

Thus, if g is indefinite then by Lemma 1.2.1-(b), $Ric = \lambda g$ at p, where $\lambda \in \mathbb{R}$, and if g is definite then by polarization identity, $Ric = \lambda g$ at p, where $\lambda \in \mathbb{R}$. □

Remark 1.2.3. Note that if (M, g) is a Riemannian manifold (that is, $\nu = 0, n$) then since R_z is diagonalizable for every $z \in S(M)$, (M, g) is Osserman at $p \in M$ if and only if the eigenvalues (counting with multiplicities) of R_z are independent of $z \in S(M)$.

In what remains of this section, we will construct examples of semi-Riemannian manifolds which are Osserman at a given point. Since the notion of algebraic curvature maps plays an essential role (cf. Remark 1.2.5), we begin by recalling some basic facts about such maps. Let V be a vector space. A quadrilinear map $F : V \times V \times V \times V \to \mathbb{R}$ is called an *algebraic curvature map* if it satisfies

$$\begin{aligned}
F(x, y, z, w) &= -F(y, x, z, w) = -F(x, y, w, z), \\
F(x, y, z, w) &= F(z, w, x, y), \\
F(x, y, z, w) &+ F(y, z, x, w) + F(z, x, y, w) = 0,
\end{aligned}$$

for all $x, y, z, w \in V$. Furthermore, if $\langle , \rangle$ is an inner product on V then the *algebraic curvature tensor* $\tilde{F} : V \times V \times V \to V$ of F with respect to $\langle , \rangle$ is defined for each $x, y, z \in V$ by $\langle \tilde{F}(x, y)z, w \rangle = F(x, y, z, w)$, where $w \in V$. Also a trilinear map $\tilde{F} : V \times V \times V \to V$ is called an *algebraic curvature tensor* with respect to an inner product $\langle , \rangle$ on V if $F : V \times V \times V \times V \to \mathbb{R}$ defined by $F(x, y, z, w) = \langle \tilde{F}(x, y)z, w \rangle$ is an algebraic curvature map, where $x, y, z, w \in V$.

Let $(V, \langle , \rangle)$ be an inner product space and $\tilde{F} : V \times V \times V \to V$ be an algebraic curvature tensor on V with respect to $\langle , \rangle$. Then the *Jacobi operator* $\tilde{F}_z : z^\perp \to z^\perp$ of $\tilde{F}$ for a nonnull unit $z \in V$ is defined by $\tilde{F}_z x = \tilde{F}(x, z)z$ and $\tilde{F}$ is called *Osserman* if the characteristic polynomial of $\tilde{F}_z$ is independent of either a timelike or spacelike unit $z \in V$. (Note that Theorem 1.2.1 also applies to algebraic curvature tensors.) Basic examples of Osserman algebraic curvature tensors are as follows:

Example 1.2.1. Let $(V, \langle , \rangle)$ be an inner product space and define an algebraic curvature tensor $R^0 : V \times V \times V \to V$ by

$$R^0(x, y)z = \langle y, z \rangle x - \langle x, z \rangle y.$$

Note that R^0 is Osserman with Jacobi operator $R_z^0 = \langle z, z \rangle id$, where z is a nonnull unit vector in V.

Example 1.2.2. A *complex structure* on a vector space V is a linear map $J : V \to V$ satisfying $J^2 = -id$. An inner product $\langle , \rangle$ on (V, J) is said to be

Hermitian if $\langle x, Jy \rangle + \langle Jx, y \rangle = 0$ for all $x, y \in V$ and the triplet $(V, J, \langle, \rangle)$ is called a *Hermitian inner product space*. Furthermore, define an algebraic curvature tensor $R^J : V \times V \times V \to V$ by

$$R^J(x, y)z = \langle Jx, z \rangle Jy - \langle Jy, z \rangle Jx + 2\langle Jx, y \rangle Jz.$$

Note that R^J is Osserman with Jacobi operator

$$R_z^J = \begin{cases} -3\langle z, z \rangle id & \text{on} \quad span\{Jz\}, \\ 0 & \text{on} \quad (span\{Jz\})^\perp \cap z^\perp, \end{cases}$$

where z is a nonnull unit vector in V.

Example 1.2.3. A *product structure* on a vector space V is a linear map $J : V \to V$ satisfying $J^2 = id$. It induces a decomposition of V into $V = V_{(+)} \oplus V_{(-)}$, where $V_{(\pm)}$ are the eigenspaces of J corresponding to eigenvalues ± 1. Conversely, if $V = V_{(+)} \oplus V_{(-)}$ is a direct sum decomposition of V then the linear map $J : V \to V$ defined by $J = \pi_{(+)} - \pi_{(-)}$ is a product structure on V, where $\pi_{(\pm)} : V \to V_{(\pm)}$ are the projections. For the special case, $dim V_{(+)} = dim V_{(-)}$, the product structure J on V is called a *paracomplex* structure on V.

An inner product $\langle, \rangle$ on a paracomplex vector space (V, J) is called *para-Hermitian* if $\langle x, Jy \rangle + \langle Jx, y \rangle = 0$ and the triplet $(V, J, \langle, \rangle)$ is called a *para-Hermitian inner product space*. Furthermore, define an algebraic curvature tensor $R^J : V \times V \times V \to V$ by

$$R^J(x, y)z = \langle Jx, z \rangle Jy - \langle Jy, z \rangle Jx + 2\langle Jx, y \rangle Jz.$$

Note that R^J is Osserman with Jacobi operator

$$R_z^J = \begin{cases} 3\langle z, z \rangle id & \text{on} \quad span\{Jz\}, \\ 0 & \text{on} \quad (span\{Jz\})^\perp \cap z^\perp, \end{cases}$$

where z is a nonnull unit vector in V.

Remark 1.2.4. It has been recently shown [52], [72], [71] that any algebraic curvature map can be expressed as a linear combination of algebraic curvature maps $R^\phi(x, y, z, v) = \phi(y, z)\phi(x, w) - \phi(x, z)\phi(y, w)$ defined by some symmetric bilinear forms ϕ.

Equivalently, any algebraic curvature map can be expressed as a linear combination of algebraic curvature maps $R^\Omega(x, y, z, v) = \Omega(y, z)\Omega(x, v) - \Omega(x, z)\Omega(y, v) - 2\Omega(x, y)\Omega(z, v)$ defined by some skew-symmetric bilinear forms Ω.

Using the Osserman algebraic curvature tensors given in Examples 1.2.1 and 1.2.2, a large family of Osserman algebraic curvature tensors can be constructed by considering Clifford module structures. Due to the important relationship between the Osserman algebraic curvature tensors and Clifford module structures, we recall here their definition and refer to Theorem 2.1.1 for necessary and sufficient conditions for their existence.

Definition 1.2.3. *Let V be an n-dimensional vector space. A real Cliff(ν)-module structure C on V, where $\nu \leq n$ is a collection of linear maps J_i on V with a set of generators $\{J_1, \ldots, J_\nu\}$ such that $J_i J_j + J_j J_i = -2\delta_{ij}$ for $i, j = 1, \ldots, \nu$.*

(That is, C determines an anti-commutative family of complex structures on V.) Note here the existence of J_i-Hermitian inner products with respect to all complex structures in a Cliff(ν)-module structure. Denote such an inner product by $\langle, \rangle$. Then one has,

Theorem 1.2.2. [66] *Suppose there is a Cliff(ν)-module structure on $\mathbb{R}^n$ and consider a set of generators $\{J_1, \ldots, J_\nu\}$ such that $J_i J_j + J_j J_i = -2\delta_{ij}$. If λ_0, $\ldots$, λ_ν are arbitrary real numbers, then the trilinear map $R : V \times V \times V \to V$ defined by*

$$R = \lambda_0 R^0 + \frac{1}{3}\sum_{i=1}^{\nu}(\lambda_i - \lambda_0)R^{J_i} \tag{1.1}$$

is an Osserman algebraic curvature tensor with

$$R_x J_i x = \lambda_i J_i x, \qquad R_x y = \lambda_0 y,$$

where x, y are nonnull unit vectors on $\mathbb{R}^n$ with y orthogonal to $\{x, J_1 x, \ldots, J_\nu x\}$.

Remark 1.2.5. Note that, for any of the algebraic curvature maps R defined above, one may construct (semi)-Riemannian metrics whose curvature tensor coincides with R at a given point. For this, let F be an algebraic curvature map on $\mathbb{R}^n$ and put $F_{ijkl} = F(e_i, e_j, e_k, e_l)$, where $\{e_s\}$ is an orthonormal basis on $\mathbb{R}^n$. Next, define a Riemannian metric tensor on the unit ball $B \subset \mathbb{R}^n$ centered at the origin by

$$g = \sum_{i,j}\left\{\delta_{ij} + \sum_{k,l}F_{ijkl}x^k x^l\right\}dx^i dx^j + O(x^3).$$

Now, the theory of normal coordinates shows that the curvature tensor of g coincides with F at the origin. By following this procedure, examples of Riemannian manifolds which are Osserman at a given point, yet whose curvature tensors do not correspond to a rank-one symmetric space, are constructed by Gilkey in [66] using the algebraic curvature maps defined in Theorem 1.2.2.

Remark 1.2.6. It is worth to empasize here the different roles played by the eigenvalue structure (Osserman property) and the Jordan form (Jordan-Osserman property) of the Jacobi operators (cf. Definition 4.2.1). Indeed, there are many examples of Osserman manifolds which are not Jordan-Osserman (see section 4.1) as well as algebraic curvature tensors which are Osserman but not Jordan-Osserman (Remark 5.1.1). Furthermore, it is now possible to exhibit examples of Osserman algebraic curvature tensors which are spacelike Jordan-Osserman but not timelike Jordan-Osserman. Following [73], let $\{e_1^-, \ldots, e_p^-, e_1^+, \ldots, e_q^+\}$ be an orthonormal basis of $(V, \langle, \rangle)$. Suppose $p > q$ and q even $(q = 2\bar{q})$ and define a linear map Φ by

$$\Phi(e_{2i-1}^+) = e_{2i}^- + e_{2i}^+, \qquad \Phi(e_{2i}^+) = -e_{2i-1}^- - e_{2i-1}^+, \qquad 1 \le i \le \bar{q}$$

$$\Phi(e_{2i-1}^-) = -e_{2i}^- - e_{2i}^+, \qquad \Phi(e_{2i}^-) = e_{2i-1}^- + e_{2i-1}^+, \qquad 1 \le i \le \bar{q}$$

$$\Phi(e_i^-) = 0, \qquad\qquad\quad , \qquad\qquad\qquad\qquad\qquad q \le i \le p.$$

Then Φ is skew-adjoint and $\Phi^2 = 0$ with ker $\Phi = \{e_1^+ + e_1^-, \ldots, e_q^+ + e_q^-, e_{q+1}^-, \ldots, e_p^-\}$. Since ker Φ contains no spacelike vectors it follows that R^Ω defined in Remark 1.2.4, where $\Omega(\cdot, \cdot) = \langle \cdot, \Phi(\cdot) \rangle$, is spacelike Jordan-Osserman. However, a simple calculation of $R_{e_1^-}^\Omega$ and $R_{e_p^-}^\Omega$ shows that R^Ω is not timelike Jordan-Osserman. (This also shows that Theorem 1.2.1 cannot be extended to Jordan-Osserman manifolds).

1.3 Semi-Riemannian Pointwise and Global Osserman Manifolds

In this section we generalize the Osserman condition to the whole manifold in two ways by giving the definitions of pointwise and global Osserman manifolds.

Definition 1.3.1. *Let (M, g) be a semi-Riemannian manifold. Then (M, g) is called pointwise Osserman if (M, g) is Osserman at each $p \in M$.*

Remark 1.3.1. Note that if (M, g) is a semi-Riemannian pointwise Osserman manifold then by Proposition 1.2.1, (M, g) is Einstein at each $p \in M$. Hence, if M is connected and $dim M \ge 3$ then, by Schur Lemma, (M, g) is an Einstein manifold, that is, $Ric = \lambda g$ on M, where $\lambda \in \mathbb{R}$.

Definition 1.3.2. *Let (M, g) be a semi-Riemannian manifold. Then (M, g) is called globally Osserman if the characteristic polynomial of R_z is independent of $z \in S^-(M)$ or $z \in S^+(M)$.*

Throughout this book, we also call a global Osserman condition, for convenience, Osserman condition whenever there is no ambiguity.

Remark 1.3.2. Let (M, g) be a semi-Riemannian globally Osserman manifold. Note that the characteristic polynomial of R_z is independent of $z \in S^-(M)$ if and only if the characteristic polynomial of R_z is independent of $z \in S^+(M)$.

(See the proof of Theorem 1.2.1.) Also recall that the characteristic polynomial of R_z for all $z \in S^-(M)$ may be different from the characteristic polynomial of R_z for all $z \in S^+(M)$.

Remark 1.3.3. A semi-Riemannian manifold (M, g) is said to be *locally isotropic* if for each point $p \in M$ and each $x, y \in T_pM$ with $g(x, x) = g(y, y)$, there is a local isometry of (M, g) of a neighborhood of p which fixes p and exchanges x and y. (Cf. [142].) Clearly, a locally isotropic semi-Riemannian manifold is Osserman.

Next we analyze the relation between the pointwise and global Osserman conditions. Let (M, g) be a semi-Riemannian manifold. Associated to the Jacobi operators, there are functions f_k defined by

$$f_k(p, z) = g(z, z)^k \mathrm{trace} R_z^{(k)}, \qquad k = 1, 2, 3, \ldots \qquad (1.2)$$

where $p \in M$, $z \in S_p(M)$ and $R_z^{(k)}$ is the kth power of the Jacobi operator R_z. Among these functions, f_1 and f_2 have a special significance. Indeed, a semi-Riemannian manifold (M, g) is *Einstein* if $f_1(p, z)$ is constant on $S(M)$, and is called 2-*stein* if f_1 and f_2 are independent of $z \in S_p(M)$ at each $p \in M$.

Note that (M, g) is pointwise Osserman if and only if the functions f_k depend only on the point p for each k. Also the global Osserman condition is equivalent to the constancy of functions f_k on $S(M)$ for each k.

The following result about the relation between the pointwise and global Osserman conditions is proved in [78] for Riemannian manifolds. Here, we present its semi-Riemannian version which is essentially obtained by following the same steps.

Theorem 1.3.1. *Let (M, g) be a connected semi-Riemannian pointwise Osserman manifold such that,*

(i) the Jacobi operators have only one eigenvalue and dim $M \geq 3$, or
(ii) the Jacobi operators have exactly two distinct eigenvalues, which are either complex or, real with constant multiplicities, at every $p \in M$ and dim $M > 4$.

Then (M, g) is globally Osserman.

Before proving this theorem, we need a technical result on the constancy of the functions f_1 and f_2 as follows.

Lemma 1.3.1. *Let (M, g) be a connected semi-Riemannian pointwise Osserman manifold.*

a) If $dim M = n \geq 3$ then f_1 is constant on M.
b) If $dim M = n > 4$ then f_2 is constant on M.

Proof. (a) Since trace $R_z = Ric(z, z)$ for all $z \in S(M)$, we have

$$Ric(z, z) = f_1(p)g(z, z) \tag{1.3}$$

for all $z \in TM$. Hence since $n \geq 3$, the function f_1 is constant on M.

(b) Let $p \in M$ and $\{e_1, \ldots, e_n\}$ be an orthonormal basis for T_pM. Then

$$f_2(p)g(z, z)^2 = traceR_z^{(2)} = \sum_{i,j=1}^{n} R(e_i, z, z, e_j)^2 \varepsilon_{e_i} \varepsilon_{e_j} \tag{1.4}$$

for any $z \in S_p(M)$, where $R(e_i, z, z, e_j) = g(R(e_i, z)z, e_j)$ and $\varepsilon_{e_i} = g(e_i, e_i)$ for $i = 1, 2, \ldots, n$. Therefore, if $x, y \in T_pM$ and $x + \sigma y$, where $\sigma = \pm 1$, then (1.4) shows that

$$f_2(p)g(x + \sigma y, x + \sigma y)^2 = \sum_{i,j=1}^{n} R(e_i, x + \sigma y, x + \sigma y, e_j)^2 \varepsilon_{e_i} \varepsilon_{e_j},$$

and

$$f_2(p) \{g(x, x) + g(y, y) + 2\sigma g(x, y)\}^2$$

$$= \sum_{i,j=1}^{n} \{R(e_i, x, x, e_j) + R(e_i, y, y, e_j) + \sigma R(e_i, x, y, e_j)$$

$$+ \sigma R(e_i, y, x, e_j)\}^2 \varepsilon_{e_i} \varepsilon_{e_j}.$$

Now, after linearizing the expression above and adding those corresponding to $\sigma = \pm 1$, one has

$$f_2(p) \{g(x, x)^2 + g(y, y)^2 + 4g(x, y)^2 + 2g(x, x)g(y, y)\}$$

$$= \sum_{i,j=1}^{n} \{R(e_i, x, x, e_j)^2 + R(e_i, y, y, e_j)^2 + R(e_i, x, y, e_j)^2 + R(e_j, x, y, e_i)^2$$

$$+ 2R(e_i, x, x, e_j)R(e_i, y, y, e_j) + 2R(e_i, x, y, e_j)R(e_j, x, y, e_i)\} \varepsilon_{e_i} \varepsilon_{e_j}.$$

Once more, applying (1.4) to x and y and using $\sum_{i,j=1}^{n} R(e_j, x, y, e_i)^2 \varepsilon_{e_i} \varepsilon_{e_j} = \sum_{i,j=1}^{n} R(e_i, x, y, e_j)^2 \varepsilon_{e_i} \varepsilon_{e_j}$, it follows that

$$f_2(p) \{g(x, x)g(y, y) + 2g(x, y)^2\} = \sum_{i,j=1}^{n} \{R(e_i, x, x, e_j)R(e_i, y, y, e_j)$$

$$+ R(e_i, x, y, e_j)R(e_j, x, y, e_i) + R(e_i, x, y, e_j)^2\} \varepsilon_{e_i} \varepsilon_{e_j}.$$

Putting $y = e_k$ in the above expression, multiplying by ε_{e_k} and adding for $k = 1, \ldots, n$, we obtain

$$f_2(p)\sum_{k=1}^{n}\left\{g(x,x)\varepsilon_{e_k}+2g(x,e_k)^2\right\}\varepsilon_{e_k}=\sum_{i,j,k=1}^{n}\left\{R(e_i,x,x,e_j)R(e_i,e_k,e_k,e_j)\right.$$

$$\left.+R(e_i,x,e_k,e_j)R(e_j,x,e_k,e_i)+R(e_i,x,e_k,e_j)^2\right\}\varepsilon_{e_i}\varepsilon_{e_j}\varepsilon_{e_k}.$$

$$(1.5)$$

Moreover, since

$$\sum_{i,j,k=1}^{n}R(e_i,x,x,e_j)R(e_i,e_k,e_k,e_j)\varepsilon_{e_i}\varepsilon_{e_j}\varepsilon_{e_k}$$

$$=\sum_{i,j=1}^{n}R(e_i,x,x,e_j)Ric(e_i,e_j)\varepsilon_{e_i}\varepsilon_{e_j},$$

and

$$\sum_{k=1}^{n}\left\{g(x,x)\varepsilon_{e_k}+2g(x,e_k)^2\right\}\varepsilon_{e_k}=ng(x,x)+2g(x,x)=(n+2)g(x,x),$$

(1.5) becomes

$$f_2(p)(n+2)g(x,x)=\sum_{i,j=1}^{n}R(e_i,x,x,e_j)Ric(e_i,e_j)\varepsilon_{e_i}\varepsilon_{e_j}$$

$$+\sum_{i,j,k=1}^{n}\left\{R(e_i,x,e_k,e_j)R(e_j,x,e_k,e_i)+R(e_i,x,e_k,e_j)^2\right\}\varepsilon_{e_i}\varepsilon_{e_j}\varepsilon_{e_k}.$$

$$(1.6)$$

Now, since $dim M = n > 2$, it follows from (a) that (M,g) is Einstein with $Ric = f_1 g$, and thus

$$\sum_{i,j=1}^{n}R(e_i,x,x,e_j)Ric(e_i,e_j)\varepsilon_{e_i}\varepsilon_{e_j}=f_1 Ric(x,x)=f_1^2 g(x,x).$$

Hence (1.6) becomes

$$f_2(p)(n+2)g(x,x)-f_1^2 g(x,x)$$

$$=\sum_{i,j,k=1}^{n}\left\{R(x,e_i,e_j,e_k)R(x,e_j,e_i,e_k)+R(x,e_i,e_j,e_k)^2\right\}\varepsilon_{e_i}\varepsilon_{e_j}\varepsilon_{e_k}.$$

$$(1.7)$$

To simplify (1.7), let Ω be the bilinear form defined by

$$\Omega(t,w)=\sum_{i,j,k=1}^{n}R(t,e_i,e_j,e_k)R(w,e_i,e_j,e_k)\varepsilon_{e_i}\varepsilon_{e_j}\varepsilon_{e_k}.\qquad(1.8)$$

Now, by the first Bianchi identity, we have

$$\sum_{i,j,k=1}^{n} R(x,e_i,e_j,e_k)R(x,e_j,e_i,e_k)\varepsilon_{e_i}\varepsilon_{e_j}\varepsilon_{e_k}$$

$$= \sum_{i,j,k=1}^{n} R(x,e_i,e_j,e_k)\left\{-R(e_j,e_i,x,e_k) - R(e_i,x,e_j,e_k)\right\}\varepsilon_{e_i}\varepsilon_{e_j}\varepsilon_{e_k}$$

$$= -\sum_{i,j,k=1}^{n} R(x,e_i,e_k,e_j)R(x,e_k,e_i,e_j)\varepsilon_{e_i}\varepsilon_{e_j}\varepsilon_{e_k}$$

$$+ \sum_{i,j,k=1}^{n} R(x,e_i,e_j,e_k)^2\varepsilon_{e_i}\varepsilon_{e_j}\varepsilon_{e_k}$$

$$= -\sum_{i,j,k=1}^{n} R(x,e_i,e_k,e_j)R(x,e_k,e_i,e_j)\varepsilon_{e_i}\varepsilon_{e_j}\varepsilon_{e_k} + \Omega(x,x) ,$$

and thus

$$\sum_{i,j,k=1}^{n} R(x,e_i,e_j,e_k)R(x,e_j,e_i,e_k)\varepsilon_{e_i}\varepsilon_{e_j}\varepsilon_{e_k} = \frac{1}{2}\Omega(x,x) . \qquad (1.9)$$

Now, from (1.7), (1.8) and (1.9), it follows that

$$f_2(p)(n+2)g(x,x) - f_1^2 g(x,x) = \frac{1}{2}\Omega(x,x) + \Omega(x,x)$$

and thus

$$\Omega(x,x) = F(p)g(x,x) , \qquad (1.10)$$

where $F : M \longrightarrow \mathbb{R}$ is the function defined by

$$F(q) = \frac{2}{3}\left(f_2(q)(n+2) - f_1^2\right), \quad q \in M . \qquad (1.11)$$

Note here that, from (1.10),

$$\Omega = F(p)g , \qquad (1.12)$$

and therefore $\sum_{i=1}^{n}\Omega(e_i,e_i)\varepsilon_{e_i} = nF(p)$. On the other hand, from (1.8),

$$\sum_{i=1}^{n}\Omega(e_i,e_i)\varepsilon_{e_i} = \sum_{i,j,k,l=1}^{n} R(e_i,e_j,e_k,e_l)^2\varepsilon_{e_i}\varepsilon_{e_j}\varepsilon_{e_k}\varepsilon_{e_l},$$

and thus (1.12) yields

$$\Omega = \frac{1}{n}\|R\|^2 g , \qquad (1.13)$$

where $\|R\|^2 = \sum\limits_{i,j,k,l=1}^{n} R(e_i,e_j,e_k,e_l)^2 \varepsilon_{e_i}\varepsilon_{e_j}\varepsilon_{e_k}\varepsilon_{e_l}.$

Also by using covariant differentiation in (1.8), we obtain

$$\sum_{b=1}^{n} (\nabla_{e_b}\Omega)(e_a,e_b)\varepsilon_{e_b}$$

$$= \sum_{b=1}^{n} \left\{ \sum_{i,j,k=1}^{n} \{(\nabla_{e_b}R)(e_a,e_i,e_j,e_k)R(e_b,e_i,e_j,e_k) \right.$$

$$\left. + R(e_a,e_i,e_j,e_k)(\nabla_{e_b}R)(e_b,e_i,e_j,e_k)\} \,\varepsilon_{e_i}\varepsilon_{e_j}\varepsilon_{e_k} \right\} \varepsilon_{e_b}$$

$$= \sum_{i,j,k,b=1}^{n} (\nabla_{e_b}R)(e_a,e_i,e_j,e_k)R(e_b,e_i,e_j,e_k)\varepsilon_{e_i}\varepsilon_{e_j}\varepsilon_{e_k}\varepsilon_{e_b}$$

$$+ \sum_{i,j,k=1}^{n} \left\{ R(e_a,e_i,e_j,e_k)\varepsilon_{e_i}\varepsilon_{e_j}\varepsilon_{e_k} \left(\sum_{b=1}^{n}(\nabla_{e_b}R)(e_b,e_i,e_j,e_k)\varepsilon_{e_b} \right) \right\},$$

and, by the second Bianchi identity,

$$\sum_{b=1}^{n} (\nabla_{e_b}R)(e_b,e_i,e_j,e_k)\varepsilon_{e_b} = \sum_{b=1}^{n} (\nabla_{e_b}R)(e_j,e_k,e_b,e_i)\varepsilon_{e_b}$$

$$= -(\nabla_{e_j}Ric)(e_i,e_k) + (\nabla_{e_k}Ric)(e_i,e_j) = 0.$$

Now, since (M,g) is Einstein,

$$\sum_{b=1}^{n} (\nabla_{e_b}\Omega)(e_a,e_b)\varepsilon_{e_b}$$

$$= \sum_{i,j,k,b=1}^{n} (\nabla_{e_b}R)(e_a,e_i,e_j,e_k)R(e_b,e_i,e_j,e_k)\varepsilon_{e_i}\varepsilon_{e_j}\varepsilon_{e_k}\varepsilon_{e_b} \tag{1.14}$$

and again, by the second Bianchi identity,

$$\sum_{i,j,k,b=1}^{n} (\nabla_{e_b}R)(e_a,e_i,e_j,e_k)R(e_b,e_i,e_j,e_k)\varepsilon_{e_i}\varepsilon_{e_j}\varepsilon_{e_k}\varepsilon_{e_b}$$

$$= \sum_{i,j,k,b=1}^{n} (\nabla_{e_a}R)(e_i,e_b,e_j,e_k)R(e_i,e_b,e_j,e_k)\varepsilon_{e_i}\varepsilon_{e_j}\varepsilon_{e_k}\varepsilon_{e_b}$$

$$- \sum_{i,j,k,b=1}^{n} (\nabla_{e_i}R)(e_a,e_b,e_j,e_k)R(e_i,e_b,e_j,e_k)\varepsilon_{e_i}\varepsilon_{e_j}\varepsilon_{e_k}\varepsilon_{e_b} .$$

Thus (1.14) reduces to

$$\sum_{b=1}^{n}(\nabla_{e_b}\Omega)(e_a,e_b)\varepsilon_{e_b}$$

$$=\frac{1}{2}\sum_{i,j,k,b=1}^{n}(\nabla_{e_a}R)(e_i,e_b,e_j,e_k)R(e_i,e_b,e_j,e_k)\varepsilon_{e_i}\varepsilon_{e_j}\varepsilon_{e_k}\varepsilon_{e_b}.$$

Now, by differentiating the expression of $\|R\|^2$, we obtain

$$\nabla_{e_a}\|R\|^2 = 2\sum_{i,j,k,b=1}^{n}(\nabla_{e_a}R)(e_i,e_b,e_j,e_k)R(e_i,e_b,e_j,e_k)\varepsilon_{e_i}\varepsilon_{e_j}\varepsilon_{e_k}\varepsilon_{e_b},$$

and thus the previous expression reduces to

$$\sum_{b=1}^{n}(\nabla_{e_b}\Omega)(e_a,e_b)\varepsilon_{e_b} = \frac{1}{4}\nabla_{e_a}\|R\|^2. \tag{1.15}$$

Also, (1.13) implies that $\sum_{b=1}^{n}(\nabla_{e_b}\Omega)(e_a,e_b)\varepsilon_{e_b} = \frac{1}{n}\sum_{b=1}^{n}g(e_a,e_b)\varepsilon_{e_b}\nabla_{e_b}\|R\|^2$, and hence

$$\sum_{b=1}^{n}(\nabla_{e_b}\Omega)(e_a,e_b)\varepsilon_{e_b} = \frac{1}{n}\nabla_{e_a}\|R\|^2. \tag{1.16}$$

Thus, since $dim M = n > 4$, (1.15) and (1.16) imply that $\nabla_{e_a}\|R\|^2 = 0$ for $a = 1,\ldots,n$, and therefore $\|R\|^2$ must be constant. Now, by (1.12) and (1.13)

$$f_2(p) = \frac{1}{n+2}\left(\frac{3}{2n}\|R\|^2 + f_1^2\right),$$

and hence, f_2 is constant.   □

Now we are ready to prove the mentioned result about the relation between the pointwise and global Osserman conditions.

Proof of Theorem 1.3.1. First note that if (M,g) is timelike Osserman at $p \in M$ with single eigenvalue $\lambda(p)$ of R_z then this eigenvalue is necessarily real. Also by the proof of Theorem 1.2.1, (M,g) is spacelike Osserman at $p \in M$ with single eigenvalue $-\lambda(p)$ of R_z. Then since $Ric(z,z) = traceR_z = (n-1)\lambda(p)g(z,z)$ for all $z \in S_p(M)$ at each $p \in M$ and $dim M = n \geq 3$, it follows that λ is constant on M and hence (M,g) is globally Osserman.

Now suppose that (M,g) is timelike Osserman at every $p \in M$ with two distinct complex eigenvalues of R_z. Then since the coefficients of the characteristic polynomial of R_z are real, these eigenvalues are complex conjugate to each other with the same multiplicities, say, $\omega(p) = \alpha(p) + i\beta(p)$ and

$\bar{\omega}(p) = \alpha(p) - i\beta(p)$ at each $p \in M$. Then by Lemma 1.3.1, since f_1 and f_2 are constant on M and, as in the above case, $trace R_z = (n-1)\alpha(p)g(z,z)$, and $trace R_z^2 = (n-1)(\alpha(p)^2 - \beta(p)^2)$ for every $z \in S_p(M)$, it follows that α and β are constant on M and hence, (M,g) is globally Osserman.

Finally suppose (M,g) is timelike Osserman at every $p \in M$ with two distinct real eigenvalues of R_z, say, $\lambda(p)$ and $\mu(p)$ with constant multiplicities m_λ and m_μ, respectively. Then by Lemma 1.3.1, since f_1 and f_2 are constant on M and, as in the first case, $trace R_z = (\lambda(p)m_\lambda + \mu(p)m_\mu)g(z,z)$, and $trace R_z^2 = \lambda(p)^2 m_\lambda + \mu(p)^2 m_\mu$ for every $z \in S_p(M)$, it follows that λ and μ are constant on M and hence (M,g) is globally Osserman.   □

Here note that we cannot deduce the constancy of λ and μ on M without assuming that they have constant multiplicities at every point. Because there is no reason that the multiplicity should not change from point to point.

Next we give some examples of semi-Riemannian globally Osserman manifolds as some model spaces. The necessary background for the following examples may be found in [101], [114], [146]. Also, examples of strictly pointwise Osserman manifolds are presented below (cf. Remark 1.3.4.)

Example 1.3.1. A semi-Riemannian manifold (M^n, g) of signature (ν, η) is called a *real space form* if (M,g) is of constant sectional curvature. Hence if (M,g) is a real space form then the curvature tensor of (M,g) is given by

$$R(x,y)z = cR^0(x,y)z,$$

where $x,y,z \in T_pM$ and $c \in \mathbb{R}$. Then the Jacobi operator of $z \in S(M)$ is given by

$$R_z = cg(z,z)\,id.$$

Thus the characteristic polynomial of R_z is $f_z(t) = (t - cg(z,z))^{n-1}$ for all $z \in S(M)$ and hence a real space form is globally Osserman. (Note that the eigenvalues of the Jacobi operators change sign from timelike to spacelike vectors (cf. Remark 1.2.2).) A complete and simply connected real space form is isometric to either of the

$$P_s^n(\mathbb{R}) = SO^s(n+1)/SO^s(n), \quad \mathbb{R}_s^n \text{, or} \quad H_s^n(\mathbb{R}) = SO^{s+1}(n+1)/SO^s(n)$$

according to the sectional curvature being positive, negative or zero (see [142].)

Example 1.3.2. Let (M,J) be an almost complex manifold with almost complex structure J (i.e., J is a (1,1)-tensor field on M satisfying $J^2 = -id$.) A semi-Riemannian metric tensor g of signature $(2\nu, 2\eta)$ is said to be Hermitian if $g(JX,Y) + g(X,JY) = 0$ for all $X,Y \in \Gamma TM$. Also (M,g,J) is said to be a *Kähler manifold* if J is a complex structure and the 2-form

$\Omega(X,Y) = g(X, JY)$ is closed. This couple of conditions can be equivalently described by $\nabla J = 0$, where ∇ is the Levi-Civita connection of g.

A plane P is called *holomorphic* if it remains invariant by the complex structure ($JP \subseteq P$), and the *holomorphic sectional curvature* is defined to be the restriction of the sectional curvature to nondegenerate holomorphic planes. A Kähler manifold (M, g, J) is called a *complex space form* if (M, g, J) is of constant holomorphic sectional curvature. Hence, the curvature tensor of (M, g, J) is given by [4],

$$R(x,y)z = \frac{c}{4} \left[R^0(x,y)z - R^J(x,y)z \right],$$

where $x, y, z \in T_pM$ and $c \in \mathbb{R}$. Then the Jacobi operator of $z \in S(M)$ is given by

$$R_z = \begin{cases} cg(z,z)\,id & \text{on} \quad span\{Jz\}, \\ \frac{c}{4}g(z,z)\,id & \text{on} \quad (span\{z, Jz\})^\perp \cap z^\perp. \end{cases}$$

Thus the characteristic polynomial of R_z is $f_z(t) = (t - cg(z,z))(t - \frac{c}{4}g(z,z))^{n-2}$ for all $z \in S(M)$ and hence a complex space form is globally Osserman. The model spaces of nonzero constant holomorphic sectional curvature are given by the symmetric spaces

$$P_s^n(\mathbb{C}) = SU^s(n+1)/U^s(n) \quad \text{and} \quad H_s^n(\mathbb{C}) = SU^{s+1}(n+1)/U^s(n)$$

(see [142].)

Remark 1.3.4. Generalizing the form of the curvature tensor of a complex space form, an almost Hermitian manifold (M, g, J) is called a *generalized complex space form* if its curvature tensor satisfies

$$R(x,y)z = fR^0(x,y)z + hR^J(x,y)z,$$

where $f, h : M \to \mathbb{R}$ are smooth functions. Generalized complex space forms are pointwise Osserman manifolds with at most two distinct eigenvalues. It is shown in [140] that generalized complex space forms are complex space forms if dim $M \geq 6$ (which also follows from Theorem 1.3.1.) Yet the existence of 4-dimensional generalized complex space forms which are not complex space forms is shown in [113].

Note that a generalized complex space form which is not a complex space form is pointwise Osserman but not globally Osserman, since f and h are nonconstant functions on M. We will further investigate the semi-Riemannian pointwise Osserman manifolds in coming chapters as well as the special role played by generalized complex space forms.

Example 1.3.3. An *almost quaternionic manifold* is a manifold M equipped with a 3-dimensional vector bundle $\mathbb{Q}$ of $(1,1)$-tensor fields on M such that there exists a local basis $\{J_1, J_2, J_3\}$ for $\mathbb{Q}$ satisfying $J_i^2 = -id$, $i = 1, 2, 3$, and $J_i J_j = J_k$, where (i, j, k) is a cyclic permutation of $(1, 2, 3)$. Such a local basis $\{J_1, J_2, J_3\}$ is called a *canonical local basis* for $\mathbb{Q}$ and $\mathbb{Q}$ is referred as an *almost quaternionic structure* on M. A semi-Riemannian metric tensor g of signature $(4\nu, 4\eta)$ is said to be *adapted* to the almost quaternionic structure $\mathbb{Q}$ if $g(\phi X, Y) + g(X, \phi Y) = 0$ for all $\phi \in \mathbb{Q}$ and $X, Y \in \Gamma TM$.

Let $(M, g, \mathbb{Q})$ be an almost quaternionic manifold and $\{J_1, J_2, J_3\}$ be a canonical local basis for $\mathbb{Q}$. Then for each $i = 1, 2, 3$, $\Phi_i(X, Y) = g(X, J_i Y)$, where $X, Y \in \Gamma TM$, is a locally defined two-form such that $\Omega = \Phi_1 \wedge \Phi_1 + \Phi_2 \wedge \Phi_2 + \Phi_3 \wedge \Phi_3$ gives rise to a globally defined 4-form on M. A quaternionic metric structure $(g, \mathbb{Q})$ is said to be *Kählerian* if Ω is parallel (or equivalently, if $\mathbb{Q}$ is parallel) with respect to the Levi-Civita connection ∇ of g (cf. [83], [128], [118].)

Let $(M, g, \mathbb{Q})$ be a quaternionic Kähler manifold. Then any vector $x \in T_p M$ determines a 4-dimensional subspace $\mathbb{Q}(x) = span\{x, J_1 x, J_2 x, J_3 x\}$ which remains invariant under the action of the quaternionic structure. We call it the $\mathbb{Q}$-section determined by x. If the sectional curvature of planes in $\mathbb{Q}(x)$ is a constant $c(x)$, where $x \in T_p M$ is nonnull, we call this constant $c(x)$ the *quaternionic sectional curvature* of (M, g) with respect to x.

A quaternionic Kähler manifold $(M, g, \mathbb{Q})$ is called a *quaternionic space form* if $(M, g, \mathbb{Q})$ is of constant quaternionic sectional curvature. Hence its curvature tensor is given by

$$R(x, y)z = \frac{c}{4}\left[R^0(x, y)z - \sum_{i=1}^{3} R^{J_i}(x, y)z\right]$$

where $x, y, z \in T_p M$, $c \in \mathbb{R}$ and $\{J_1, J_2, J_3\}$ is a canonical local basis for $\mathbb{Q}$. Then the Jacobi operator with respect to $z \in S(M)$ is given by

$$R_z = \begin{cases} cg(z, z)\, id & \text{on} \quad span\{J_1 z, J_2 z, J_3 z\}, \\ \frac{c}{4}g(z, z)\, id & \text{on} \quad (span\{J_1 z, J_2 z, J_3 z\})^\perp \cap z^\perp. \end{cases}$$

Thus the characteristic polynomial of R_z is $f_z(t) = (t - cg(z, z))^3(t - \frac{c}{4}g(z, z))^{n-4}$ for all $z \in S(M)$ and hence a quaternionic space form is globally Osserman. A nonflat quaternionic space form is isometric to any of the model spaces

$$P_s^n(\mathbb{Q}) = Sp^s(n+1)/Sp^s(n) \times Sp(1), \quad H_s^n(\mathbb{Q}) = Sp^{s+1}(n+1)/Sp^s(n) \times Sp(1)$$

(see [142].)

In addition to the well-known examples of semi-Riemannian Osserman manifolds described above, there are some other examples which have no Riemannian analog. However, they may be considered as a kind of real version of both complex and quaternionic manifolds.

Example 1.3.4. A *para-Kähler manifold* is a symplectic manifold locally diffeomorphic to a product of Lagrangian submanifolds. Such a product induces a decomposition of the tangent bundle TM into a Whitney sum of Lagrangian subbundles L and L', that is, $TM = L \oplus L'$. By generalizing this definition, an *almost para-Hermitian manifold* is defined to be an almost symplectic manifold (M, Ω) whose tangent bundle splits into a Whitney sum of Lagrangian subbundles. This definition implies that the $(1,1)$-tensor field J defined by $J = \pi_L - \pi_{L'}$ is an almost paracomplex structure, that is, $J^2 = id$, on M such that $\Omega(JX, JY) = -\Omega(X, Y)$ for all $X, Y \in \Gamma TM$, where π_L and π'_L are the projections of TM onto L and L', respectively. The 2-form Ω induces a nondegenerate $(0, 2)$-tensor field g on M defined by $g(X, Y) = \Omega(X, JY)$, where $X, Y \in \Gamma TM$. Now, by using the relation between the almost paracomplex and the almost symplectic structures on M, it follows that g defines a semi-Riemannian metric tensor of signature (n, n) on M and moreover, $g(JX, Y) + g(X, JY) = 0$, where $X, Y \in \Gamma TM$. The special significance of the para-Kähler condition is equivalently stated in terms of the parallelizability of the paracomplex structure with respect to the Levi-Civita conection of g, that is, $\nabla J = 0$ [42].

A plane P is called *paraholomorphic* if it is left invariant by the action of paracomplex structure J, that is, $JP \subseteq P$. Now the *paraholomorphic sectional curvature H* is defined by the restriction of the sectional curvature to paraholomorphic nondegenerate planes. A para-Kähler manifold (M, g, J) is called a *paracomplex space form* if (M, g, J) is of constant paraholomorphic sectional curvature. Hence, the curvature tensor of (M, g, J) is given by [54]

$$R(x, y)z = \frac{c}{4}[R^0(x, y)z + R^J(x, y)z],$$

where $x, y, z \in T_pM$ and $c \in \mathbb{R}$. Then the Jacobi operator of $z \in S(M)$ is given by

$$R_z = \begin{cases} cg(z, z)\, id & \text{on} \quad span\{Jz\}, \\ \frac{c}{4}g(z, z)\, id & \text{on} \quad (span\{z, Jz\})^\perp \cap z^\perp. \end{cases}$$

Thus the characteristic polynomial of R_z is $f_z(t) = (t - cg(z, z))(t - \frac{c}{4}g(z, z))^{n-2}$ for all $z \in S(M)$ and hence a paracomplex space form is globally Osserman. Nonflat complete and simply connected paracomplex space forms are isometric to the symmetric spaces $SL(n, \mathbb{R})/SL(n - 1, \mathbb{R}) \times \mathbb{R}$ (see [62].)

Example 1.3.5. A *paraquaternionic structure* $\mathbb{Q}$ on a manifold M is defined to be a 3-dimensional bundle of $(1,1)$-tensor fields on M such that there exists a local basis $\{J_1, J_2, J_3\}$ for $\mathbb{Q}$ satisfying

$$J_i^2 = \varepsilon_i id, \qquad J_1 J_2 = -J_2 J_1 = J_3.$$

where $\varepsilon_1 = \varepsilon_2 = -\varepsilon_3 = 1$. A semi-Riemannian metric tensor g on M of signature $(2n, 2n)$ is said to be *adapted* to the paraquaternionic structure $\mathbb{Q}$ if it satisfies $g(J_iX, Y) + g(X, J_iY) = 0$ for all $X, Y \in \Gamma TM$ and any local basis for $\mathbb{Q}$. Moreover, $(M, g, \mathbb{Q})$ is called a *paraquaternionic Kähler manifold* if the bundle $\mathbb{Q}$ is parallel with respect to the Levi-Civita connection of g (cf. [14], [62].)

Let $(M, g, \mathbb{Q})$ be a paraquaternionic Kähler manifold. Then any vector $x \in T_pM$ determines a 4-dimensional subspace $\mathbb{Q}(x) = span\{x, J_1x, J_2x, J_3x\}$ which remains invariant under the action of the paraquaternionic structure. We call it the $\mathbb{Q}$-section determined by x. Note that the restriction of the metric tensor g to any $\mathbb{Q}$-section is indefinite of signature $(2, 2)$ or totally degenerate, where the latter case occurs if and only if the $\mathbb{Q}$-section is generated by a null vector. If the sectional curvature of nondegenerate planes in $\mathbb{Q}(x)$ is a constant $c(x)$, where $x \in T_pM$ is nonnull, we call this constant $c(x)$ the *paraquaternionic sectional curvature* of (M, g) with respect to x.

A paraquaternionic Kähler manifold $(M, g, \mathbb{Q})$ is called a *paraquaternionic space form* if $(M, g, \mathbb{Q})$ is of constant paraquaternionic sectional curvature. Hence, its curvature tensor satisfies

$$R(x, y)z = \frac{c}{4}\left[R^0(x, y)z + \sum_{i=1}^{3} \varepsilon_i R^{J_i}(x, y)z\right]$$

where $\varepsilon_1 = \varepsilon_2 = -\varepsilon_3 = 1$, $x, y, z \in T_pM$ and $c \in \mathbb{R}$. Then the Jacobi operator with respect to $z \in S(M)$ is given by

$$R_z = \begin{cases} cg(z, z)\, id & \text{on} \quad span\{J_1z, J_2z, J_3z\} \\ \frac{c}{4}g(z, z)\, id & \text{on} \quad (span\{J_1z, J_2z, J_3z\})^{\perp} \cap z^{\perp}. \end{cases}$$

Thus the characteristic polynomial of R_z is $f_z(t) = (t - cg(z, z))^3(t - \frac{c}{4}g(z, z))^{n-4}$ for all $z \in S(M)$ and hence a paraquaternionic space form is globally Osserman.

Complete and simply connected nonflat paraquaternionic space forms are isometric to the symmetric spaces $Sp(n; \mathbb{R})/Sp(1; \mathbb{R}) \times Sp(n - 1; \mathbb{R})$.

2. The Osserman Conjecture in Riemannian Geometry

Let (M, g) be an isotropic Riemannian manifold. Then it is a two-point homogeneous space, i.e. the group of local isometries acts transitively on the unit sphere bundle of (M, g) and hence (M, g) is Osserman. The lack of other examples led Osserman to conjecture that the converse might also be true [116], which was proved by Chi [39] in many special cases.

In this chapter we concentrate our attention to Riemannian Osserman manifolds. First of all, it is important to note that although there have been important progresses toward the complete proof of the conjecture, it is still an open problem at the present for the Riemannian case. In fact Chi [39] proved the conjecture for Riemannian manifolds of dimensions $n \neq 4m$, $m \geq 1$. Also a study of Gilkey in [66] motivated himself together with Swann and Vanhecke to introduce the pointwise Osserman condition in [78], where by using this weaker condition they also obtained a positive answer to the Osserman conjecture when the Riemannian manifold has dimension $n \neq 4m$, $m \geq 1$.

In section 1, we first show that if the manifold has dimension $n \neq 4m$, $m > 1$, then the Jacobi operators have at most two distinct eigenvalues and the pointwise Osserman condition is equivalent to the global Osserman condition. In §2.1.1 we give a proof of the Osserman conjecture for dimension $2m + 1$ and in §2.1.2 we give its proof for dimension $4m + 2$, by employing the methods in [78]. The special case of dimension 4 is treated in §2.1.3.

Section 2 is devoted to the analysis of the Osserman condition with some additional assumptions. In §2.2.1, we deal with the case of Kähler and quaternionic-Kähler Riemannian Osserman manifolds. In §2.2.2, we collect the study of Chi when the Jacobi operator of the curvature tensor of a Riemannian Osserman manifold is assumed to satisfy two natural additional assumptions. In §2.2.3 we analyze the Osserman condition for homogeneous Riemannian manifolds of negative sectional curvature.

Throughout this chapter, we only consider Riemannian manifolds unless otherwise stated.

2.1 Partial Solution to the Osserman Conjecture

In this section we give a proof of the Osserman conjecture for Riemannian manifolds of dimension $n \neq 4m$, $m > 1$. This result was first obtained by Chi (cf. [39]) considering the global Osserman condition, and it was generalized later by Gilkey, Swann and Vanhecke [78] when the pointwise version of the Osserman conjecture is considered with the exception of dimension 4. Here we follow the ideas in paper [78]. The authors suggested in [78] that a proof of the Osserman conjecture could be given by using a two-step process. The first step should consist of the characterization of all possible Osserman curvature tensors which may be realized at a single point of the Riemannian manifold. The second step would be based on the use of the second Bianchi identity, which could lead to show that such spaces must be locally symmetric.

We use the above two-step process to give a proof of the Osserman conjecture when the dimension is $n \neq 4m$, $m \geq 1$. To begin, we look at an interesting relation of certain topological conditions to the existence of Clifford structures, which are also related to the existence of certain distributions on TS^{m-1}.

Theorem 2.1.1. [135] *Let* $n = 2^r \cdot n_0$, *with odd* n_0.

(i) *An n-dimensional vector space V admits a Cliff(ν)-module structure if and only if $\nu \leq \nu(r)$,*

(ii) *TS^{n-1} admits a k-dimensional distribution for $2k \leq n - 1$ if and only if $k \leq \nu(r)$,*

where $\nu(r)$ is given by $\nu(i + 4) = \nu(i) + 8$ and $\nu(i) = 2^i - 1$ for $i = 0, 1, 2, 3$.

The following two theorems are immediate consequences of Theorem 2.1.1 which will be used in this and subsequent chapters.

Theorem 2.1.2. *The n-dimensional sphere S^n does not admit a continuous k-dimensional distribution if n is even and $1 \leq k \leq n - 1$.*

Theorem 2.1.3. *The $(4m + 1)$-dimensional sphere S^{4m+1} does not admit a continuous k-dimensional distribution if $2 \leq k \leq n - 2$, where $n = 4m + 1$.*

2.1.1 Odd-Dimensional Riemannian Pointwise Osserman Manifolds

In this subsection, we give a proof of the Osserman conjecture in Riemannian geometry for dimension $n = 2m + 1$.

Theorem 2.1.4. *If (M, g) is a connected Riemannian pointwise Osserman manifold of dimension $n = 2m + 1$ then (M, g) is a real space form. Hence, it is either flat or locally a rank-one symmetric space.*

Proof. For each $x \in S_p(M)$, identify $T_x S_p(M)$ with $x^\perp$, and let $\lambda \in \mathbb{R}$ be an eigenvalue of R_x for all $x \in S_p(M)$. Now consider the bundle homomorphism T_λ defined by $T_\lambda = R_x - \lambda id$ on $S_p(M) \cong S^{2m}$. Then by Theorem 2.1.2, $ker T_\lambda = TS_p(M)$ and hence, the eigenvalues of R_x are equal, that is, $R_x = \lambda id$ for all $x \in S_p(M)$. Thus, for all orthogonal $x, y \in S_p(M)$, $g(R(x, y)y, x) = \lambda$, that is, (M, g) is of constant sectional curvature at p. Then it follows from Schur Lemma that (M, g) is of constant sectional curvature. $\qquad\square$

2.1.2 $(4m + 2)$-Dimensional Riemannian Pointwise Osserman Manifolds

In this subsection we provide a positive answer to the Osserman conjecture in Riemannian geometry when the manifold is of dimension $4m + 2$.

Let (M, g) be a $(4m + 2)$-dimensional Riemannian pointwise Osserman manifold. As in the proof of the previous theorem, for each $x \in S_p(M)$, identify $T_x S_p(M)$ with $x^\perp$, and let $\lambda_1 \in \mathbb{R}$ be an eigenvalue of R_x for all $x \in S_p(M)$. Now consider the bundle homomorphism T_{λ_1} defined by $T_{\lambda_1} = R_x - \lambda_1 id$ on $S_p(M) \cong S^{4m+1}$. Then, it follows from Theorem 2.1.3 that the dimension of $\ker T_{\lambda_1}$ is either 1 or $4m$. Next, let λ_i be another eigenvalue of R_x. Since λ_i must be of multiplicity 1 or $4m$ as above, it immediately follows that the multiplicity of λ_i is $4m$, since otherwise $ker(R_x - \lambda_1 id) \oplus Ker(R_x - \lambda_i id)$ defines a 2-dimensional distribution on S^{4m+1}, in contradiction to Theorem 2.1.2. Therefore, the Jacobi operators R_x have either one eigenvalue or exactly two distinct eigenvalues, with multiplicities 1 and $4m$, at each $p \in M$. Thus, as a consequence of Theorem 1.3.1 and the argument above, it immediately follows that any connected $(4m + 2)$-dimensional Riemannian pointwise Osserman manifold that is not of constant sectional curvature at each $p \in M$ is necessarily a global Osserman manifold with exactly two distinct eigenvalues of multiplicities 1 and $4m$. For, if the Jacobi operators R_x at $p \in M$ have only one eigenvalue then the sectional curvature is constant at $p \in M$. Therefore, we may assume that the Jacobi operators have two distinct constant eigenvalues λ and μ with multiplicities 1 and $4m$, respectively.

Now, we show that, if the Jacobi operators have exactly two distinct constant eigenvalues with multiplicity 1 and $4m$ on a Riemannian manifold, then the manifold corresponds to a complex space form. The proof is divided into two steps. The first step corresponds to the explicit determination of the form of the curvature tensor at each point (cf. Lemma 2.1.4) and it is based on the pointwise manipulation of the identities of an algebraic curvature tensor. In the second step, we show that the manifold is Kählerian and the holomorphic sectional curvature is constant. This will be achieved by making an extensive use of the second Bianchi identity. We begin with the following observation.

Let (M, g) be a Riemannian globally Osserman manifold with exactly two distinct eigenvalues λ and μ of the Jacobi operators R_x with multiplicities 1 and $4m$, respectively. Then

$$y \in Ker(R_x - \lambda id) \Leftrightarrow R_x y = \lambda y \Leftrightarrow R(y, x, x, y) = \lambda$$
$$\Leftrightarrow R(x, y, y, x) = \lambda \Leftrightarrow R_y x = \lambda x$$
$$\Leftrightarrow x \in Ker(R_y - \lambda id) \, ,$$

and hence, y is a unit vector in ker $(R_x - \lambda id)$ if and only if $x \in$ ker $(R_y - \lambda id)$, where $R(x, y, z, v) = g(R(x, y)z, v)$. This is a special case of a more general duality property shown by Rakić in [124]. Now consider the line bundle over $S(M)$ corresponding to the eigenvalue λ of multiplicity 1 and assume that M is contractible (note that this assumption does not cause any loss of generality.) Then, since any line bundle over $S^{n-1} \cong S_p M$ is trivial (see [135]), there exists a smooth global unit section ξ of the line subbundle ker T_λ of $S(M)$. Now define a map J on $S(M)$ by $Jx = \xi_x$, where $\xi_x \in$ ker $(R_x - \lambda id)$. Then we have the following properties of J:

Lemma 2.1.1. *a) J satisfies $J^2(X) = -X$ and $J(-X) = -J(X)$ for all $X \in \Gamma S(M)$.*
b) The extension of J to TM is linear.

Proof. a) Put $V_X = span\{X, JX\}$ and note that, if $Y \in \Gamma S(M)$ then, $Y \perp V_X$ if and only if $X \perp V_Y$. Also, if U is a unit vector field in V_X then $U = \cos(\theta)X + \sin(\theta)JX$. Defining $Z(U) = -\sin(\theta)X + \cos(\theta)JX$, the sectional curvature of V_X is given by

$$R(U, Z(U), Z(U), U) = R(X, JX, JX, X) = \lambda.$$

Thus it follows that $R_U Z(U) = \lambda Z(U)$ and $Z(U) = \pm JU$. Hence, by continuity, the sign is independent of the angle θ, and thus $J^2(X) = Z^2(X) = -X$ and $J(-X) = \pm Z(-X) = \mp Z(X) = -J(X)$.

b) Extend J to TM by defining $J(aX) = aJ(X)$, where $a \in \mathbb{R}$. Then note that it suffices to show that $J(X + Y) = JX + JY$ for $Y \perp V_X$ since for arbitrary sections A and B of TM, one can rescale A to be unit length and by writing $B = A' + B'$, where $A' \in V_A$ and $B' \perp V_A$, it follows that

$$J(A + B) = J(A + A' + B') = J(A + A') + JB'$$
$$= JA + JA' + JB' = JA + J(A' + B') = JA + JB \, .$$

Thus we only need to show that

$$J(\cos(\theta)X + \sin(\theta)Y) = \cos(\theta)JX + \sin(\theta)JY \tag{2.1}$$

for unit vector fields X and Y with $Y \perp V_X$. For this, let $\alpha = \cos(\theta)$ and $\beta = \sin(\theta)$, and define $J' = \varepsilon J$ for some $\varepsilon = \pm 1$. Let $A_\theta = \alpha X + \beta Y$ and $B_\theta = \alpha JX + \beta J'Y$. Note here, if we assume that, for some choice of ε,

$$b_\theta = R(B_\theta, A_\theta, A_\theta, B_\theta) = \lambda \tag{2.2}$$

for all angles θ, then $B_\theta = \pm J(A_\theta)$ and by continuity, the sign does not change. Taking $\theta = 0$, it follows that $B_\theta = J(A_\theta)$ and for $\theta = \pi/2$, one gets $J'Y = JY$ and $\varepsilon = 1$. Thus (2.1) holds.

To prove (2.2), first note that $R(\cdot, \cdot, \cdot, \cdot)$ vanishes whenever three entries of R are in $\{X, JX\}$ and one in $\{Y, J'Y\}$ or vice versa. Thus, expanding b_θ in terms of X, JX, Y and $J'Y$, the coefficients of $\alpha^3\beta$ and $\alpha\beta^3$ are equal to 0. Moreover, $R(J'Y, X, X, J'Y) = R(JX, Y, Y, JX) = \mu$ and hence

$$b_\theta = (\alpha^4 + \beta^4)\lambda + 2\alpha^2\beta^2\{R(JX, X, Y, J'Y) + \mu + R(J'Y, X, Y, JX)\}. \quad (2.3)$$

Then to prove (2.2), it suffices to show that

$$R(JX, X, Y, J'Y) + \mu + R(J'Y, X, Y, JX) = \lambda,$$

that is,

$$R(JX, X, Y, J'Y) + R(J'Y, X, Y, JX) = \lambda - \mu. \quad (2.4)$$

To complete the proof, we show that (2.4) holds. This is a consequence of the following results.

Lemma 2.1.2. (1) $R(U, S)T = -R(U, T)S$ when $S \perp T$ and $S, T \in \Gamma V_U^\perp$,
(2) $R(S, T)U = 0$ when $S \perp V_T$ and $S, T \in \Gamma V_U^\perp$,
(3) $2R(X, Y, J'Y, JX) = R(JX, X, Y, J'Y)$,
(4) $2R(J'Y, X, Y, JX) = R(JX, X, Y, J'Y)$.

Proof. Putting $W = \cos(\phi)S + \sin(\phi)T$, we have U orthogonal to V_W, and thus $R(U, W)W = \mu U$. Expanding this identity, we get

$$(\cos^2(\phi) + \sin^2(\phi))\mu U + \cos(\phi)\sin(\phi)\{R(U, S)T + R(U, T)S\} = \mu U,$$

which proves (1). For (2), using (1) and the first Bianchi identity, we obtain

$$R(S, T)U = -R(T, U)S - R(U, S)T = R(T, S)U + R(S, U)T = -2R(S, T)U,$$

which proves (2). For (3), note that (1) implies that

$$\begin{aligned}R(X, Y, J'Y, JX) &= -R(Y, J'Y, X, JX) - R(J'Y, X, Y, JX) \\ &= R(JX, X, Y, J'Y) - R(J'Y, JX, X, Y).\end{aligned}$$

Finally (4) follows from

$$\begin{aligned}R(J'Y, X, Y, JX) &= -R(Y, JX, X, J'Y) = R(Y, X, JX, J'Y) \\ &= \frac{1}{2}R(JX, X, Y, J'Y).\end{aligned}$$

$\square$

Lemma 2.1.3. The curvature tensor of (M, g) satisfies

$$R(JX, X, Y, J'Y) = \pm\frac{2(\lambda - \mu)}{3}.$$

Proof. Let $A = A_{\pi/4} = (X + Y)/\sqrt{2}$, $B = B_{\pi/4} = (JX + J'Y)/\sqrt{2}$, and define

$$C = \frac{X - Y}{\sqrt{2}}, \qquad D = \frac{JX - J'Y}{\sqrt{2}}.$$

Then $\{A, B, C, D\}$ is an orthonormal basis for $V_X \oplus V_Y$ and B is an eigenvector of R_A, as we show next.

If Z is orthogonal to $V_X \oplus V_Y$, then

$$2\sqrt{2}R(B, A, A, Z) = R(JX, X, X, Z) + R(JX, X, Y, Z) + R(J'Y, X, X, Z)$$
$$+R(J'Y, X, Y, Z) + R(JX, Y, X, Z) + R(JX, Y, Y, Z)$$
$$+R(J'Y, Y, X, Z) + R(J'Y, Y, Y, Z).$$

Since $R(JX, X)X = \lambda JX$, the first term vanishes and $R(JX, X, Y, Z) = -R(Y, Z, X, JX) = 0$. Thus the second term also vanishes. Similarly the remaining five terms vanish and hence $R(B, A, A, Z) = 0$. Now note that

$$4R(B, A, A, C) = -2R(JX + J'Y, X + Y, X, Y)$$
$$= -2\{R(JX, X, X, Y) + R(J'Y, X, X, Y)$$
$$+R(JX, Y, X, Y) + R(J'Y, Y, X, Y)\} = 0,$$

and

$$4R(B, A, A, D) = R(JX, X, X, JX) - R(JX, X, X, J'Y)$$
$$+R(JX, X, Y, JX) - R(JX, X, Y, J'Y) + R(J'Y, X, X, JX)$$
$$-R(J'Y, X, X, J'Y) + R(J'Y, X, Y, JX) - R(J'Y, X, Y, J'Y)$$
$$+R(JX, Y, X, JX) - R(JX, Y, X, J'Y) + R(JX, Y, Y, JX)$$
$$-R(JX, Y, Y, J'Y) + R(J'Y, Y, X, JX) - R(J'Y, Y, X, J'Y)$$
$$+R(J'Y, Y, Y, JX) - R(J'Y, Y, Y, J'Y)$$
$$= \lambda - R(JX, X, Y, J'Y) - \mu + R(J'Y, X, Y, JX)$$
$$-R(JX, Y, X, J'Y) + \mu + R(J'Y, Y, X, JX) - \lambda = 0.$$

Thus, $R(B, A)A = b'B$ for some b', as claimed. As in (2.3), we have

$$b' = \frac{1}{2}(\lambda + \mu) + \frac{3}{4}R(JX, X, Y, J'Y),$$

and hence the claim follows. □

Now, the above results immediately enable us to complete the proof of Lemma 2.1.1-(b). Indeed, by choosing $J' = \varepsilon J$, it follows that $R(JX, X, Y, J'Y) = 2(\lambda - \mu)/3$, and thus (2.4) holds. □

Hence we obtained the expression of the curvature tensor of (M, g) as follows.

Lemma 2.1.4. *Let (M, g) be a $(4m+2)$-dimensional Riemannian Osserman manifold which is not of constant sectional curvature at each point $p \in M$. Then the curvature tensor of (M, g) is given by*

$$R = \lambda R^0 + \mu' R^J,$$

where $\mu' = (\lambda - \mu)/3 \neq 0$ and, λ and μ are the eigenvalues of the Jacobi operators with multiplicities 1 and $4m$, respectively.

The previous lemma shows that (M, g) is a generalized complex space form and the final characterization of $(4m + 2)$-dimensional Riemannian Osserman manifolds is obtained as follows.

Theorem 2.1.5. *Let (M, g) be a connected $(4m + 2)$-dimensional Riemannian pointwise Osserman manifold which is not of constant sectional curvature at each $p \in M$. Then (M, g) is a complex space form, and hence locally a rank-one symmetric space.*

Proof. Since $dim M = 4m + 2$, it follows from Theorem 1.3.1 and Lemma 2.1.4 that (M, g) is a real space form or a generalized complex space form. Hence, if (M, g) is not of constant curvature, then the curvature tensor of (M, g) takes the form

$$R = f R^0 + h R^J, \tag{2.5}$$

where f and h are constant functions on M with $h \neq 0$. Hence

$$\nabla_W R = f \nabla_W R^0 + h \nabla_W R^J,$$

where $W \in \Gamma TM$. Also note that $\nabla R^0 = 0$ and

$(\nabla_W R^J)(X, Y)Z$

$\quad = -g((\nabla_W J)Y, Z)JX + g((\nabla_W J)X, Z)JY + 2g((\nabla_W J)X, Y)JZ$

$\quad -g(JY, Z)(\nabla_W J)X + g(JX, Z)(\nabla_W J)Y + 2g(JX, Y)(\nabla_W J)Z .$

Now, let Y be a unit vector field orthogonal to span$\{X, JX\}$. By the second Bianchi identity, $(\nabla_Y R)(X, JY)X + (\nabla_X R)(JY, Y)X + (\nabla_{JY} R)(Y, X)X = 0$, and hence

$$0 = -2(\nabla_X J)X + \{g(X, (\nabla_X J)Y) - g(X, (\nabla_Y J)X)\}Y$$

$$+\{3g(X, (\nabla_{JY} J)Y) - g(X, (\nabla_Y J)JY)$$

$$+2g(Y, (\nabla_X J)JY) + 2g(JY, (\nabla_Y J)X)\}JX$$

$$+\{g(X, (\nabla_X J)JY) - g(X, (\nabla_{JY} J)X)\}JY .$$

Thus, we obtain

$$2g((\nabla_X J)X, Y) - g(X, (\nabla_X J)Y) + g(X, (\nabla_Y J)X) = 0 ,$$

and, since $g(X, (\nabla_Y J)X) = 0$ and $g(X, (\nabla_X J)Y) = -g((\nabla_X J)X, Y)$, it follows that $(\nabla_X J)X$ is a section of span$\{X, JX\}$. But since $(\nabla_X J)X$ is also orthogonal to X and JX, we conclude that $(\nabla_X J)X = 0$. Therefore (M, g, J) is a nearly Kähler manifold. Further note from the expression of the curvature tensor in (2.5) that the holomorphic sectional curvature is constant. A nearly Kähler manifold of constant holomorphic sectional curvature is known to be Kähler or locally isometric to the 6-dimensional sphere with nearly Kähler structure induced by the product of the Cayley numbers [80]. Therefore, since (M, g) is not a space form, it must be Kähler, and thus locally isometric to a complex space form.   □

Motivated by the previous result, manifolds with curvature tensors given by (1.1) are investigated in [78]. We include here some of those results, which are needed in Chapter 3. Suppose we have a Cliff(ν)-structure with generators $J_1, \ldots, J_\nu$ as given in Theorem 1.2.2 and assume that the curvature tensor R is of the form

$$R = \lambda_0 R^0 + \sum_{i=1}^{\nu} \lambda_i R^{J_i}$$

for some functions $\lambda_0, \ldots, \lambda_\nu$.

Lemma 2.1.5. *Let $n = dimM$ and assume the Cliff(ν)-structure on (M, g) satisfies one of the following six conditions:*

a) $\nu > 9$,

b) $n > 4\nu$,

c) $\nu = 3$, $n > 8$ and TM decomposes as a direct sum of isomorphic irreducible Cliff(3)-modules (or, equivalently, $J_3 = \pm J_1 J_2$),

d) $\nu = 5$ and $n > 16$,

e) $\nu = 6$ and $n > 16$,

f) $\nu = 7$, $n > 16$ and TM decomposes as a direct sum of isomorphic irreducible Cliff(7)-modules.

Then we have

(1) for all $X \in \Gamma TM$ and $i = 1, \ldots, \nu$, the vector field $\lambda_i (\nabla_X J_i)X$ is in the linear span of $\{J_j X, j \neq i\}$

(2) the functions $\lambda_0, \ldots, \lambda_\nu$ are constant.

Lemma 2.1.6. *Let $n = dimM$ and assume the Cliff(ν)-structure on (M, g) satisfies either (a') $\nu > 8$ or (b') $n > 2\nu$. Suppose that λ_0 is constant. Then we have*

(1) for all $X \in \Gamma TM$ and $i = 1, \ldots, \nu$, the vector field $\lambda_i (\nabla_X J_i)X$ is in the orthogonal complement of $J_i X$ in the linear span of $\{J_j X, J_j J_i X; i \neq j\}$,

(2) the functions $\lambda_1, \ldots, \lambda_\nu$ are constant.

The previous results are mainly applied to the case of pointwise Osserman manifolds admitting a compatible Cliff(2)-structure with generators J_1, J_2. Defining $J_3 = J_1 J_2$ and considering the more general case arising from the resulting Cliff(3)-structure, the curvature tensor may be written as

$$R = \lambda_0 R^0 + \lambda_1 R^{J_1} + \lambda_2 R^{J_2} + \lambda_3 R^{J_3},$$

where $\lambda_0, \lambda_1, \lambda_2, \lambda_3$ are real functions on M.

Theorem 2.1.6. *Let (M, g) be a Riemannian pointwise Osserman manifold with a quaternionic structure as above. If*

a) $\dim M \geq 12$, or
b) $\dim M \geq 8$ and λ_0 is constant,

then (M, g) is either flat or locally isomorphic to a rank-one symmetric space of real, complex or quaternionic type.

2.1.3 Four-Dimensional Riemannian Osserman Manifolds

Four-dimensional manifolds have a special significance when considering the Osserman problem. First of all, note that neither Theorem 1.3.1 nor Theorem 2.1.4 nor Theorem 2.1.5 cover 4-dimensional manifolds. Although an analogous result to Theorem 2.1.5 can be proven in dimension 4, the main difference of which from the cases considered by now is the existence of pointwise Osserman manifolds which are not globally Osserman.

First note that the Osserman conjecture holds in dimension 4, which is shown in [39]

Theorem 2.1.7. *If (M, g) is a 4-dimensional Riemannian globally Osserman manifold then it is either flat or locally a rank-one symmetric space.*

We will not provide a proof of this theorem here, since we will give a proof of the corresponding result for semi-Riemannian manifolds in Chapter 4.

In the remaining part of this subsection, we focus on the existence of examples of 4-dimensional Riemannian pointwise Osserman manifolds which are not globally Osserman. First note the following relation between the class of pointwise Osserman manifolds and the class of self-dual manifolds. (Again we do not go into details about the proof because we will deal with analogous results for general semi-Riemannian manifolds in Chapter 4.) Moreover, note that a Riemannian manifold (M, g) is pointwise Osserman if and only if it is k-stein for all k (i.e., at each point $p \in M$, $\mathrm{trace} R_x^{(k)}$ is independent of the unit $x \in T_p M$ for all k.) Since any 4-dimensional 2-stein manifold is 3-stein [10] the equivalence between (a) and (c) in the following theorem is clear.

Theorem 2.1.8. [78] *Let (M, g) be a 4-dimensional Riemannian manifold. Then, the following conditions are equivalent:*

a) (M, g) is pointwise Osserman.

b) Locally there is a choice of orientation for M for which the metric tensor g is self-dual and Einstein.

c) (M, g) is 2-stein.

This result enables us to give several examples, as it is shown in the following result. We follow the discussion in [78]

Corollary 2.1.1. [78] *There exist 4-dimensional Riemannian pointwise Osserman manifolds which are not globally Osserman and hence they are not locally isometric to rank-one symmetric spaces. There are also examples where the function $\|R\|^2$ is not constant.*

Theorem 2.1.7 solves the pointwise Osserman conjecture in dimension 4. By taking Theorem 2.1.8 into consideration, the first part of the above corollary is established by giving an example of a self-dual Einstein manifold of dimension 4 which is not locally symmetric. A compact example is obtained by taking (M, g) as a $K3$-surface. By Yau's proof of the Calabi conjecture, this carries a hyperKähler metric tensor and, in particular, it is self-dual and Ricci-flat. (See for example [11].) Local examples with nonzero scalar curvature are obtained by taking an open set in one of the quaternionic Kähler orbifolds constructed in [55] or [56]. Further hyperKähler examples with many possibilities for the eigenvalues are found in [97].

An example where $\|R\|^2$ is nonconstant is given by the Calabi metric tensor on $T^*\mathbb{C}P(1)$. (See [33], [34], [65].) This is a complete hyperKähler metric tensor which has an action of $U(2)$ such that the central $U(1)$ fixes a complex structure I and rotates J and K. Thus at any given point, two eigenvalues of the Jacobi operators are equal and these eigenvalues μ_1 and μ_2 generate all the symmetric functions of the eigenvalues of the Jacobi operator. Now, μ_1 is constant since (M, g) is Einstein, in fact, Ricci-flat. However, μ_2 can not be constant; since otherwise (M, g) would be globally Osserman and hence locally symmetric. Thus, μ_2 and hence $\|R\|^2$ are nonconstant functions for this metric tensor.

A second example where $\|R\|^2$ is nonconstant arises from the work of Olszak [113], which we next treat in more detail. We recall that an almost Hermitian manifold (M, g, J) is called a generalized complex space form if its curvature tensor R is of the form

$$R = fR^0 + hR^J , \tag{2.6}$$

where f and h are smooth functions on M. It is shown in [140] that generalized complex space forms are indeed complex space forms, provided that $dim M \geq 6$. (This is also derived from the results in the previous subsection, since generalized complex space forms are pointwise Osserman manifolds.) Further note that 4-dimensional generalized complex space forms are Hermitian on the open subset where $h \neq 0$ and $f + h$ is constant. Moreover, if either f or h is constant then (M, g, J) is a 4-dimensional complex space form. In [113] Olszak constructed generalized complex space forms, where h is nonconstant, by means of certain conformal deformations of Bochner-flat Kählerian manifolds with nonconstant scalar curvature. It is also shown that any generalized complex space form for which h is nonzero at each point and nonconstant can be obtained only in that way. Next we include these results.

Theorem 2.1.9. [113] *Let* $(M, \tilde{g}, J)$ *be a Bochner-flat Kähler manifold of dimension* 4. *In addition, assume that the scalar curvature* $\tilde{\tau}$ *of* $\tilde{g}$ *is nonzero everywhere on* M *and nonconstant. Let* $g = e^{\sigma} \tilde{g}$, *where* $\sigma = -\log(C\tilde{\tau}^2)$, $C(= const.) > 0$. *Then the Hermitian manifold* (M, g, J) *is a generalized complex space form for which the function* $h \neq 0$ *everywhere on* M *and* $h \neq const.$

Theorem 2.1.10. [113] *Let* (M, g, J) *be a generalized complex space form of dimension* 4 *for which the function* $h \neq 0$ *at each point of* M *and* $h \neq const.$ *Let* $\tilde{g} = e^{-\sigma} g$, *where* $\sigma = -(1/3) \log(C_1 h^2)$, $C_1(= const.) > 0$. *Then* $(M, \tilde{g}, J)$ *is a Bochner-flat Kähler manifold and* $\sigma = -\log(C\tilde{\tau}^2)$, $C(= const.) > 0$.

Remark 2.1.1. Examples of self-dual, and hence Bochner-flat Kähler manifolds of dimension 4 with nonconstant scalar curvature, can be found in [46].

2.2 A Solution to the Osserman Conjecture with Additional Assumptions

Although the Osserman conjecture in Riemannian geometry remains open, it has been proven to be true under additional hypotheses. Here we pay attention to the cases of complex and quaternionic Kähler manifolds, homogeneous manifolds, and to a special case when the Jacobi operators are assumed to have a simple form.

2.2.1 Kähler and Quaternionic Kähler Manifolds

The sectional curvature of a Kähler manifold (M, g, J) is constant if and only if (M, g) is flat. Therefore, the holomorphic sectional curvature is introduced as being the restriction of the sectional curvature to holomorphic planes. At each point $p \in M$, the holomorphic sectional curvature H may be viewed as a real-valued function defined on the unit sphere S^{2n-1} and therefore it is necessarily bounded. Hence, if the holomorphic sectional curvature is nonnegative, there is no loss of generality in assuming (M, g, J) to be δ-*holomorphically pinched*, that is, there is some $\delta > 0$ such that $\delta \leq H(x) \leq 1$, where $H(x) = \kappa(span\{x, Jx\})$ denotes the curvature of the holomorphic plane $span\{x, Jx\}$. Now one has the following:

Lemma 2.2.1. [12] *Let x and y be orthonormal vectors tangent to a δ-holomorphically pinched Kähler manifold (M, g, J) at a point $p \in M$. Then, if $g(x, Jy) = \cos \theta$,*

$$\frac{1}{4} \left[3(1 + \cos^2 \theta)\delta - 2 \right] \leq \kappa(span\{x, y\}) \leq 1 - \frac{3\delta(1 - \cos^2 \theta)}{4}.$$

Here observe that the sectional curvature κ achives its maximum at a plane $span\{x, y\}$ if and only if $g(x, Jy) = 1$, that is, if and only if $span\{x, y\}$ is a holomorphic plane. Hence, we have the following [39]:

Theorem 2.2.1. *Let (M, g, J) be a Kähler manifold of nonnegative or nonpositive sectional curvature. Then (M, g, J) is pointwise Osserman if and only if (M, g, J) is a complex space form.*

Proof. Since (M, g, J) is assumed to be pointwise Osserman, let $\lambda_1(p)$, $\ldots$, $\lambda_{2n-1}(p)$ denote the eigenvalues of the Jacobi operators with respect to $x \in S_p(M)$, where $\lambda_1(p)$ is the maximum of the eigenvalues at $p \in M$. Since $\lambda_1(p)$ indicates the maximum value of the sectional curvature at p, it follows from Lemma 2.2.1 that the holomorphic sectional curvature $H(x) = \lambda_1(p)$ is constant at each $p \in M$. Then, the Schur lemma shows that (M, g, J) is a Kähler manifold of constant holomorphic sectional curvature, that is, a complex space form.

The case of nonpositive curvature is obtained similarly by considering the minimum of sectional curvatures. □

Remark 2.2.1. Note that the above result can also be obtained under the assumption of nonnegativity or nonpositivity of the holomorphic sectional curvatures. Moreover, an extension of Theorem 2.2.1 has been recently proven in [24] for some classes of almost Hermitian manifolds.

Next consider the case of quaternionic Kähler manifolds. Following the same ideas as above, let $(M, g, \mathbb{Q})$ be a quaternionic Kähler manifold and let $\{J_1, J_2, J_3\}$ be a local canonical basis for the quaternionic structure $\mathbb{Q}$.

Lemma 2.2.2. [40] *Let $(M, g, \mathbb{Q})$ be a quaternionic Kähler manifold. Let $x, y \in T_pM$ be orthonormal vectors and put $\cos \beta_i = g(J_i x, y)$, $i = 1, 2, 3$. Then*

$$24\kappa(span\{x, y\}) = -12\frac{Sc}{4k(k+2)} \left[1 - \sum_{i=1}^{3} \cos^2 \beta_i \right]$$

$$+ \left[\sum_{i=1}^{3} 3(1 + \cos \beta_i)^2 H_i(x + J_i y) \right.$$

$$\left. + 3(1 - \cos \beta_i)^2 H_i(x - J_i y) \right],$$

where $H_i(z) = R(z, J_i z, J_i z, z)/\|z\|^4$.

Now, assume that the scalar curvature is nonnegative and let $x, y \in T_p M$ orthonormal vectors spanning a plane where the sectional curvature achives its maximum. (This must be a nonnegative δ by the assumption of the non-negativity of the scalar curvature.) Then

$$\delta = \kappa(span\{x, y\})$$
$$\leq -\frac{Sc}{4k(k+2)}\left[1 - \sum_{i=1}^{3} \cos^2 \beta_i\right] + \frac{\delta}{8}\sum_{i=1}^{3}\left[(1 + \cos \beta_i)^2 + (1 - \cos \beta_i)^2\right]$$
$$= \delta - \left[1 - \sum_{i=1}^{3} \cos^2 \beta_i\right]\left(\frac{\delta}{4} + \frac{Sc}{8k(k+2)}\right)$$

and hence $\sum_{i=1}^{3} \cos^2 \beta_i = 1$ because $\{x, y\}$ span a quaternionic plane. Therefore, the maximum of the sectional curvature of a nonflat quaternionic Kähler manifold of nonnegative scalar curvature is the quaternionic sectional curvature. (Similarly, the minimal sectional curvature of a nonflat quaternionic Kähler manifold of nonpositive scalar curvature is the quaternionic sectional curvature.) With this lemma we have the following [40].

Theorem 2.2.2. *Let $(M, g, \mathbb{Q})$ be a pointwise Osserman quaternionic Kähler manifold of nonnegative (resp., nonpositive) scalar curvature. If the maximal (resp., minimal) eigenvalue of the Jacobi operators has multiplicity three, then $(M, g, \mathbb{Q})$ is a quaternionic space form.*

Proof. First of all, note that if $dim M = 4$, then (M, g) is a space form due to the existence of an eigenvalue of the Jacobi operators with multiplicity 3. Next assume $dim M > 4$. Then, since the multiplicity of the maximal (resp., minimal) eigenvalue of the Jacobi opertors is 3, (M, g) cannot be flat at any point. Furthermore, since the maximum (resp., minimum) of the sectional curvatures is achived at a quaternionic plane, the quaternionic sectional curvature of $(M, g, \mathbb{Q})$ is constant at each point due to the fact that the multiplicity of the maximal (resp., minimal) eigenvalue is 3. This shows that the curvature tensor of $(M, g, \mathbb{Q})$ is given by Example 1.3.3 at every $p \in M$ and the claim follows from Theorem 1.3.1.   □

By Theorem 2.2.2, the Osserman conjecture is true once the multiplicity of the maximal eigenvalue of the Jacobi operators in positive scalar curvature (resp., minimal eigenvalue if the scalar curvature is negative) is three. This is proven by Chi under the condition that the second Betti number is zero.

Theorem 2.2.3. [40] *A compact simply connected quaternionic Kähler manifold $(M, g, \mathbb{Q})$ with $H^2(M, \mathbb{R}) = 0$ must be the quaternionic projective space, provided that the maximal (resp., minimal if the scalar curvature is negative) eigenvalue of the Jacobi operators depends only on the base point with a fixed constant multiplicity.*

2.2.2 Riemannian Osserman Manifolds with Two Distinct Eigenvalues

If (M, g) is a two-point homogeneous manifold of nonconstant sectional curvature, then the associated Jacobi operators have exactly two distinct eigenvalues, one with multiplicity 1, 3 or 7, corresponding to complex space forms, quaternionic space forms and the Cayley planes, respectively. This observation is the starting point in [41] where the two-point homogeneous manifolds (of nonconstant sectional curvature) are characterized by the following two conditions:

(1) The Jacobi operators R_x have exactly two distinct eigenvalues λ and μ for every $x \in S_p(M)$ and $p \in M$.

(2) Let $E_\lambda(x)$ be the linear space spanned by x and the eigenspace corresponding to the eigenvalue λ, that is,

$$E_\lambda(x) = span\{x\} \oplus ker(R_x - \lambda id) .$$

Then $E_\lambda(x) = E_\lambda(y)$ whenever $y \in E_\lambda(x)$,

Note here that the condition (2) above is satisfied by any two-point homogeneous manifold since $E_\lambda(x)$ is nothing but the tangent space to a totally geodesic sphere of curvature λ through the base point x. Further note that, if (M, g) is a complex space form, a quaternionic space form or the Cayley plane, then for any unit $x \in T_pM$, the subspace $E_\lambda(x)$ corresponds to the complex, quaternionic or Cayley subspace spanned by x, respectively.

Theorem 2.2.4. [41] *Let (M, g) be a Riemannian manifold satisfying (1) and (2) above. Then it is locally isometric to a two-point homogeneous space of nonconstant sectional curvature.*

In the remaining part of this subsection we briefly recall some steps toward the proof of this theorem. We do not consider most of the technical details and refer the reader to Chapter 5, where a corresponding result is proved for semi-Riemannian manifolds.

The first step in proving Theorem 2.2.4 is the determination of the possible multiplicities of the distinguished eigenvalue λ. This is achieved by using the necessary and sufficient conditions for the existence of a Clifford module structure in Theorem 2.1.1. Here note that the conditions (1) and (2) allow us to obtain an orthogonal decomposition of the tangent space at each point of (M, g) as

$$T_pM = E_\lambda(x_0) \oplus E_\lambda(y_0) \oplus E_\lambda(z_0) \oplus \cdots . \qquad (2.7)$$

Next, fix a subspace $E_\lambda(x_0)$ and define certain linear maps on $E_\lambda(x_0)$ as follows: Let $\{x_1, x_2, \ldots, x_\tau\}$ be an orthonormal basis for $ker(R_{x_0} - \lambda id)$, where $\dim E_\lambda(x_0) = \tau + 1$, and put

$$J_i : \xi \in E_\lambda(x_0)^\perp \mapsto J_i(\xi) = R(x_0, x_i)\xi \in E_\lambda(x_0)^\perp .$$

Note that J_i is well-defined for each $i = 1, \ldots, \tau$ and moreover, for each $y_0 \in E_\lambda(x_0)^\perp$, J_i is a linear map on $E_\lambda(y_0)$. Furthermore, it also follows that

(i) $J_i^2 = -id$,
(ii) $J_i J_k + J_k J_i = 0$ for $i \neq k$,
(iii) $J_i J_k(y) \in span\{J_1(y), \ldots, J_\tau(y)\}$,

for all $i, j \in \{1, \ldots, \tau\}$.

Hence, $\{J_1, \ldots, J_\tau\}$ defines a $\mathrm{Cliff}(\tau)$-module structure on $E_\lambda(y_0)$. Now, since $dim E_\lambda(y_0) = \tau + 1$, it follows from Theorem 2.1.1 that $\tau \in \{1, 3, 7\}$. Moreover, since $\{J_1, \ldots, J_\tau\}$ generates a $\mathrm{Cliff}(\tau)$-module structure on $E_\lambda(x_0)^\perp$, it follows that $\tau = 7$ can only occur when $dim M = 16$.

The second part of the proof of Theorem 2.2.4 can be outlined as follows. Since τ must be equal to 1, 3 or 7, $E_\lambda(y_0)$ can be equipped with a product induced by the $\mathrm{Cliff}(\tau)$-module structure as follows:

$$y_0 \cdot y_i = y_i \, ,$$
$$y_i \cdot y_k = J_i(y_k) \, , \qquad i \neq k \, .$$

This product endowes $E_\lambda(y_0)$ with a structure isomorphic to one of the associative algebras $\mathbb{C}$, $\mathbb{H}$ or the nonassociative Cayley algebra Ca, respectively.

Note that there exists a basis $\{x_0, x_1, \ldots, x_\tau, y_0, y_1, \ldots, y_\tau, z_0, z_1, \ldots, z_{n-(2\tau+2)}\}$ induced by the decomposition $T_p M = E_\lambda(x_0) \oplus E_\lambda(y_0) \oplus \cdots$, such that the components of the curvature tensor are given by

$$R(x_0, x_\alpha) y_\beta = -y_{\beta\alpha}, \qquad R(x_0, x_\alpha) z_\beta = -z_{\beta\alpha},$$

where $\alpha\beta$ denotes the standard multiplication for the three algebras. The existence of a local basis as above and the second Bianchi identity lead to the local symmetry of the manifold. Finally, the result can be obtained from an examination of the holonomy of (M, g), or directly from Theorem 2.2.5.

2.2.3 Homogeneous Manifolds of Negative Sectional Curvature

In trying to understand the geometry of Osserman manifolds, it seems natural to impose some symmetry conditions on the structure of the manifold. The first such natural condition is the local symmetric character. Independent of the Osserman problem, although motivated by their study on harmonic manifolds, Carpenter, Gray and Willmore [36] studied symmetric spaces which are Einstein and, more precisely, 2-stein. An immediate consequence of their work is the following:

Theorem 2.2.5. *A locally symmetric Riemannian manifold (M, g) is Osserman if and only if it is flat or locally rank-one symmetric.*

This provides a positive answer to the Osserman conjecture in locally symmetric setting.

Remark 2.2.2. The result in the previous theorem can be generalized to the broader class of semi-symmetric Riemannian manifolds. A Riemannian manifold is said to be semi-symmetric if, at each point of M, the Riemannian curvature tensor is the same as that of a symmetric space (which may vary from point to point.) The algebraic characterization of semi-symmetric spaces is given by the formula $R(X, Y) \cdot R = 0$, where the dot denotes derivation in the tensor algebra.

Now, based on the result of Theorem 2.2.5 the curvature tensor at each point of a semi-symmetric Riemannian Osserman manifold (M, g) is vanishing or that of a rank-one symmetric space. Since the Osserman condition ensures the constancy of the multiplicities of the eigenvalues of the Jacobi operators, it immediately follows that any Osserman semi-symmetric space is either flat or locally rank-one [25].

A natural step in generalizing the result of Theorem 2.2.5 is to consider the broader class of homogenous Einstein manifolds. It was proven in [2] that Ricci flat homogeneous Riemannian manifolds are necessarily flat (see also [6], [11].) Therefore, only those cases where $Sc \neq 0$ deserve further investigation. In what remains of this subsection we will consider the case $Sc < 0$.

All known examples of noncompact, nonflat, homogeneous Einstein Riemannian manifolds are isometric to Einstein solvmanifolds $(S, \langle, \rangle)$, where S denotes a connected and simply connected solvable Lie group and $\langle, \rangle$ is a left-invariant Einstein metric tensor on S. Then known examples can be modelled on certain metric Lie algebras $(\mathfrak{s}, \langle, \rangle)$, which we will refer to as *Lie algebras of Iwasawa type* satisfying the following conditions:

(i) The orthogonal complement $\mathfrak{a}$ of $\mathfrak{n} = [\mathfrak{s}, \mathfrak{s}]$ is abelian.
(ii) All ad_A, $A \in \mathfrak{a}$, are symmetric relative to $\langle, \rangle$ and nonzero for $A \neq 0$.
(iii) For some $A^0 \in \mathfrak{a}$, the restriction of $ad_{A^0}|_\mathfrak{n}$ is positive definite.

We call any left-invariant Einstein metric tensor modelled on a Lie algebra of Iwasawa type a *standard Einstein solvmanifold*.

Note that, for a homogeneous Einstein manifold G/K, if the sectional curvatures of G/K are nonpositive, then the isotropy subgroup K is necessarily a maximal compact subgroup of G. (Such a maximallity has been conjectured by Alekseevsky for any homogenous Einstein manifold G/K with $Sc < 0$) [1], [11]. Now, it follows from the analysis of Heber that,

Theorem 2.2.6. [82] *Let $(G/K, g)$ be a noncompact homogeneous space with K maximal and compact in G, where g is a G-invariant metric tensor. If $(G/K, g)$ is isometric to a standard Einstein solvmanifold, then any other G-invariant metric tensor on G/K is isometric to g modulo scaling.*

Therefore, in order to investigate the Osserman problem inside the class of homogeneous Einstein manifolds of nonpositive sectional curvature, we may restrict ourselves to analyze it at the Lie algebra level of any Einstein Lie algebra of Iwasawa type. This is the approach Dotti and Druetta followed in [49], where they obtained the following.

Theorem 2.2.7. [49] *If (S, g) is a solvable Lie group of Iwasawa type satisfying the Osserman condition, then (S, g) is a symmetric manifold of noncompact type of rank-one.*

Therefore, a positive answer to the above mentioned conjecture of Alekseevsky on the maximality of the isotropic subgroup also provides a positive answer to the Osserman problem in the class of homogeneous manifolds with $Sc < 0$.

The proof of Theorem 2.2.7 can be sketched as follows [49].

First of all, note that if $(\mathfrak{s}, \langle, \rangle)$ is a Lie algebra of Iwasawa type satisfying the Osserman condition, then dim $\mathfrak{a} = 1$ and (S, g) has negative sectional curvature. Next, let $\mathfrak{z}$ denote the center of $\mathfrak{n} = [\mathfrak{s}, \mathfrak{s}]$ and let $\mathfrak{b}$ be the orthogonal complement of $\mathfrak{z}$ with respect to the metric tensor $\langle, \rangle$ restricted to $\mathfrak{n}$. Thus $\mathfrak{n} = \mathfrak{z} \oplus \mathfrak{b}$ and $\mathfrak{s} = \mathfrak{n} \oplus \mathbb{R}H$, where $H \in \mathfrak{a}$ is chosen so that $\langle H, H \rangle = 1$ and all the eigenvalues of $ad_H |_{\mathfrak{n}}$ are positive. Moreover, since ad_H is symmetric, $ad_H : \mathfrak{z} \mapsto \mathfrak{z}$ and $ad_H : \mathfrak{b} \mapsto \mathfrak{b}$. Now, if $\{\lambda_i\}$ are the eigenvalues of $ad_H |_{\mathfrak{n}}$, denote by $\mathfrak{n}_{\lambda_i}$ the eigenspace associated to the eigenvalue λ_i. Hence, $\mathfrak{n}$ decomposes as an orthogonal direct sum of $\mathfrak{n}_{\lambda_i}$ which are invariant by ad_H and satisfy $[\mathfrak{n}_{\lambda_i}, \mathfrak{n}_{\lambda_j}] \subset \mathfrak{n}_{\lambda_i + \lambda_j}$, whenever $\lambda_i + \lambda_j$ is an eigenvalue. Moreover, note that the center of $\mathfrak{n}$ becomes an eigenspace of ad_H, that is, $ad_H |_{\mathfrak{z}} = \lambda id$.

Finally, the Authors in [49] show that if $(\mathfrak{s}, \langle, \rangle)$ is Osserman, then either $\mathfrak{n}$ is abelian (and then, since $ad_H |_{\mathfrak{n}} = \lambda id$, the Lie group S corresponds to a real hyperbolic space form) or otherwise, $ad_H |_{\mathfrak{n}}$ has only two eigenvalues λ and $\mu = \frac{1}{2}\lambda$.

Consequently, $ad_H |_{\mathfrak{z}} = \lambda id$ and $ad_H |_{\mathfrak{b}} = \frac{1}{2}\lambda id$ and the Lie algebra $\mathfrak{s}$ can be decomposed as $\mathfrak{s} = \mathfrak{n} \oplus \mathbb{R}H_0$, where H_0 is a unit vector in $\mathfrak{a}$ and

$$ad_{H_0} |_{\mathfrak{z}} = id, \qquad ad_{H_0} |_{\mathfrak{b}} = \frac{1}{2}id, \qquad j_Z^2 = -\langle Z, Z \rangle id \quad \text{for all } Z \in \mathfrak{z},$$

where $\langle j_Z X, Y \rangle = \langle [X, Y], Z \rangle$ for all $X, Y \in \mathfrak{b}$.

Hence $\mathfrak{s}$ is the canonical 1-dimensional solvable extension of an H-type nilpotent $\mathfrak{n}$. Since the sectional curvatures are negative, it follows from [48] that S is a symmetric space of noncompact type of rank-one.

3. Lorentzian Osserman Manifolds

A semi-Riemannian manifold (M, g) of index $\nu = 1$ is called a *Lorentzian manifold*. In Lorentzian geometry, working with timelike vectors shows certain differences from working with spacelike vectors. A major difference is that a timelike vector has an induced definite inner product, as opposed to a spacelike vector that has an induced indefinite inner product. Hence most tools of linear algebra of definite inner product spaces may be used in the orthogonal spaces of timelike vectors, such as the diagonalizability of self-adjoint linear maps defined on the orthogonal spaces of timelike vectors. In Section 3.1, we will use this distinct property of timelike vectors to prove the Osserman conjecture in Lorentzian geometry, that is, a Lorentzian global Osserman manifold is a real space form and hence, is either flat or locally rank-one symmetric. In Section 3.2, we introduce the null Osserman condition. Again in Lorentzian geometry, null vectors show closer properties to timelike vectors rather than to spacelike vectors. Although the orthogonal space to a null vector is degenerate, that is, the restriction of the metric tensor to this space is a degenerate bilinear form, by quotienting out its degenerate part, we obtain a definite inner product quotient space. This enables us to define the Jacobi operator with respect to a null vector and hence to introduce the null Osserman condition. Also in this section we give curvature characterizations of some Lorentzian null Osserman manifolds and their local decompositions as a consequence of the form of their curvature tensors.

3.1 The Complete Solution to the Osserman Problem

In the literature the solution of the Osserman problem in Lorentzian geometry was obtained by studying the timelike and spacelike cases separately (cf. [16], [58], [59].) Although the solution of the conjecture is rather easy for the timelike case, since Theorem 1.2.1 was not known at that time, the proof of the spacelike case was given by a somewhat difficult proof. Here in the light of the equivalence of timelike and spacelike Osserman conditions at a point by Theorem 1.2.1, we provide a simple proof of the Osserman conjecture in Lorentzian geometry. We also need the following well-known fact in semi-Riemannian geometry.

At each point p on a Riemannian manifold (M, g), the sectional curvature is a continuous function defined on the Grassmannian of two-planes tangent to (M, g) at p, that is, $\kappa : G_2(T_pM) \to \mathbb{R}$ is a continuous function. An immediate consequence of the compactness of $G_2(T_pM)$ is that the sectional curvature is bounded at each point. However, the sectional curvature of indefinite metrics is only defined for nondegenerate planes, that is, at each point $p \in M$, κ is defined on the open set $\tilde{G}_2(T_pM) = G_2(T_pM) \backslash \{P = span\{x, y\}$ such that $g(x, x)g(y, y) - g(x, y)^2 = 0\}$. Due to the noncompactness of $\tilde{G}_2(T_pM)$, κ is not necessarily bounded at each point.

Then, two basic problems appear in the study of the sectional curvatures of indefinite metrics, as to, when κ is bounded at a given $p \in M$ and when it is possible to continuously extend κ to the whole Grassmannian $G_2(T_pM)$. The second problem was extensively studied by Graves and Nomizu, and Dajczer and Nomizu in [79] and [45], respectively. The first problem, on the boundedness of the sectional curvature was considered by Kulkarni in [98]. We summarize the answers to both problems in the next lemma.

Lemma 3.1.1. [114] *Let (M, g) be a semi-Riemannian manifold of index $0 < \nu < n$. Then the following conditions are equivalent.*

a) *$\kappa(P)$ is constant for all nondegenerate planes P tangent to (M, g) at $p \in M$, that is, $R(x, y)z = cR^0(x, y)z$ for every $x, y, z \in T_pM$,*
b) *$R(x, y, y, x) = 0$ for all degenerate planes $P = \{x, y\}$ tangent to (M, g) at $p \in M$,*
c) *$a \leq \kappa(P)$ or $\kappa(P) \leq b$ for all nondegerate planes P tangent to (M, g) at $p \in M$, where $a, b \in \mathbb{R}$,*
d) *$a \leq \kappa(P) \leq b$ for all timelike planes P (that is, the restriction of g to P is of signature $(-+)$) tangent to (M, g) at $p \in M$,*
e) *$a \leq \kappa(P) \leq b$ for all definite planes P (that is, the restriction of g to P is either $(--)$ or $(++)$) tangent to (M, g) at $p \in M$.*

Although we are mainly interested in (d) of the above lemma, we give a proof of (c), which is due to Kulkarni. Then (d) can be obtained by similar arguments, and we refer the reader to [114, p. 229] for the proof of the remaining part of Lemma 3.1.1. Let (M, g) be a semi-Riemannian manifold and let the sectional curvature be bounded from below by a real a at a point $p \in M$. Take $x, y, z \in T_pM$ nonnull orthogonal unit vectors satisfying $g(x, x) = g(y, y) = 1$, $g(z, z) = -1$. Then, for each $\lambda \neq \pm 1$, one has

$$a \leq \kappa(span\{\lambda x + z, y\}) = \frac{\lambda^2 R(x, y, y, x) + 2\lambda R(x, y, y, z) + R(z, y, y, z)}{\lambda^2 - 1},$$

and hence

$$\lambda^2 \kappa(span\{x, y\}) + 2\lambda R(x, y, y, z) - \kappa(span\{y, z\}) \leq a(\lambda^2 - 1) \quad \text{if } |\lambda| < 1$$
$$\lambda^2 \kappa(span\{x, y\}) + 2\lambda R(x, y, y, z) - \kappa(span\{y, z\}) \geq a(\lambda^2 - 1) \quad \text{if } |\lambda| > 1.$$

Then by continuity,

$$\lambda^2 \kappa(span\{x,y\}) + 2\lambda R(x,y,y,z) - \kappa(span\{y,z\}) = 0$$

if $\lambda = \pm 1$. This implies that $R(x,y,y,z) = 0$ for all orthogonal vectors with $g(x,x) = g(y,y) = -g(z,z)$. Hence, the constancy of the sectional curvature at $p \in M$ follows from the results in [45, Thm.5] (see also Lemma 3.2.1 below.)

Now, we are ready to state the main result of this section.

Theorem 3.1.1. *Let (M,g) be a Lorentzian manifold. Then, (M,g) is Osserman at $p \in M$ if and only if (M,g) is of constant sectional curvature at $p \in M$.*

Proof. Clearly by Example 1.3.1, if (M,g) is of constant sectional curvature at $p \in M$ then (M,g) is Osserman at $p \in M$. Conversely, consider (M,g) as timelike Osserman at $p \in M$ and let $z \in S_p^-(M)$. Then since R_z is a self-adjoint linear map on $z^\perp$ and $z^\perp$ has definite induced inner product, R_z is diagonalizable. Let $c_1, \ldots, c_{n-1}$ be the eigenvalues (counting with multiplicities) of R_z with corresponding orthonormal eigenvectors $v_1, \ldots, v_{n-1}$, respectively. Now let $x = \sum_{i=1}^{n-1} a_i v_i \in z^\perp$ be a unit vector. Then

$$g(R_z x, x) = \sum_{i=1}^{n-1} a_i g(R_z v_i, x) = \sum_{i=1}^{n-1} a_i c_i g(v_i, x) = \sum_{i=1}^{n-1} a_i^2 c_i.$$

Hence,

$$|\, g(R_z x, x) \,| \leq \sum_{i=1}^{n-1} |\, a_i \,|^2 c_i \,| \leq \sum_{i=1}^{n-1} |\, c_i \,|,$$

since $|\, a_i \,|^2 \leq 1$ for each $i = 1, \ldots, n-1$. But since

$$|\, g(R_z x, x) \,| = |\, g(R(x,z)z, x) \,| = |\, \kappa(P) \,|,$$

where $P = span\{z, x\}$, it follows that

$$|\, \kappa(P) \,| \leq \sum_{i=1}^{n-1} |\, c_i \,|.$$

Now since (M,g) is timelike Osserman at $p \in M$ and $z^\perp$ has definite induced inner product, it follows from the constancy of the coefficients of the characteristic polynomial of R_z for all $z \in S_p^-(M)$ that $c_1, \ldots, c_{n-1}$ are constant for all $z \in S_p^-(M)$. Thus, for every timelike plane P tangent to (M,g) at p, we have

$$| \kappa(P) | \leq \sum_{i=1}^{n-1} | c_i | \qquad \text{(constant)}.$$

Hence by Lemma 3.1.1-(d), (M, g) is of constant sectional curvature at $p \in M$.
□

Thus, we have the solution to the Osserman problem in Lorentzian geometry.

Theorem 3.1.2. *If (M, g) is a connected $(n \geq 3)$-dimensional Lorentzian pointwise Osserman manifold, then (M, g) is a real space form, that is, (M, g) is of constant sectional curvature.*

Proof. By Theorem 3.1.2, (M, g) is of constant sectional curvature at each $p \in M$. Hence by the Schur lemma, (M, g) is of constant sectional curvature, provided that $dim M \geq 3$. □

Remark 3.1.1. Note that the classes of Osserman manifolds and isotropic manifolds coincide in the Lorentzian setting.

Remark 3.1.2. For the sake of completeness, we also give a separate proof that Lorentzian spacelike Osserman manifolds are real space forms. The proof also uses some specific features of Lorentzian geometry and therefore, it may be of interest to the reader.
Let $\{e_1, \ldots, e_n\}$ be an orthonormal basis for $T_p M$, where e_n is timelike. Further, let $R_{ijkl} = R(e_i, e_j, e_k, e_l)$ denote the components of the curvature tensor of (M, g) with respect to the basis $\{e_1, \ldots, e_n\}$ at $p \in M$. The essential point is to make a hyperbolic rotation in the first and last coordinates to create a new orthonormal basis. In fact the existence of such rotations is the crucial difference between the Lorentzian and Riemannian geometries.
Let α, β be real numbers with $\beta^2 - \alpha^2 = 1$ and define a new orthonormal frame by setting

$$x_1 = \beta e_1 + \alpha e_n \qquad x_i = e_i \ (1 < i < n) \qquad x_n = \beta e_n + \alpha e_1.$$

Now suppose (M, g) is spacelike Osserman at $p \in M$. Expand $R_{x_1} x_u = \sum_v a_{uv} x_v$ with $a_{1v} = a_{v1} = 0$ and

$$a_{ij} = a_{ji} = -R(x_1, x_i, x_1, x_j) = -\beta^2 R_{1i1j} - \alpha\beta(R_{1inj} + R_{ni1j}) - \alpha^2 R_{ninj},$$

$$a_{in} = -a_{ni} = R(x_1, x_n, x_1, x_i) = \beta R_{1n1i} - \alpha R_{n1ni},$$

$$a_{nn} = R_{1n1n}.$$

Now by using the identity $\beta^2 = \alpha^2 + 1$,

$$trace R_{x_1}^{(2)} = a_{nn}^2 + \sum_{i,j} a_{ij}^2 - 2 \sum_i a_{in}^2$$

$$= a_{nn}^2 + \sum_{i,j}(-\beta^2 R_{1i1j} - \alpha\beta(R_{1inj} + R_{ni1j}) - \alpha^2 R_{ninj})^2$$

$$-2\sum_i(\beta R_{1n1i} - \alpha R_{n1ni})^2$$

$$= a_{nn}^2 + \sum_{i,j}(\beta^4 R_{1i1j}^2 + \alpha^2\beta^2(R_{1inj} + R_{ni1j})^2 + \alpha^4 R_{ninj}^2$$

$$+2\alpha^2\beta^2 R_{1i1j}R_{ninj}) - 2\sum_i(\alpha^2 R_{n1ni}^2 + \beta^2 R_{1n1i}^2) + \mathcal{E}(\alpha,\beta)$$

$$= \alpha^4 \sum_{i,j}(R_{1i1j}^2 + (R_{1inj} + R_{ni1j})^2 + R_{ninj}^2 + 2R_{1i1j}R_{ninj})$$

$$+\alpha^2 \sum_{i,j}(2R_{1i1j}^2 + (R_{1inj} + R_{ni1j})^2 + 2R_{1i1j}R_{ninj})$$

$$-2\alpha^2 \sum_i(R_{n1ni}^2 + R_{1n1i}^2) + \mathcal{C} + \mathcal{E}(\alpha,\beta),$$

where $\mathcal{C}$ is independent of α and $\mathcal{E}(\alpha,\beta)$ is an odd function of α.

Since the eigenvalues of R_{x_1} are constant on $S_p^+(M)$, $trace R_{x_1}^{(2)}$ is independent of the choice of α and β. Therefore, the coefficient of α^4 must be zero, and thus

$$\sum_{i,j}\{(R_{1inj} + R_{ni1j})^2 + (R_{1i1j} + R_{ninj})^2\} = 0.$$

Now, expand $R_{x_n}x_u = \sum_v b_{uv}x_v$ similarly with $b_{nn} = 0$, $b_{in} = b_{ni} = 0$, $b_{1n} = b_{n1} = 0$, and

$$b_{ij} = b_{ji} = -R(x_n, x_i, x_n, x_j)$$
$$= -\beta^2 R_{ninj} - \alpha\beta(R_{1inj} + R_{ni1j}) - \alpha^2 R_{1i1j},$$
$$b_{1j} = b_{j1} = -R(x_n, x_1, x_n, x_j) = -\beta R_{n1nj} + \alpha R_{1n1j},$$
$$b_{11} = -R_{n1n1}.$$

Hence,

$$trace R_{x_n}^{(2)} = \alpha^4 \sum_{i,j}((R_{ninj}^2 + (R_{1inj} + R_{ni1j})^2 + R_{1i1j}^2 + 2R_{1i1j}R_{ninj})$$

$$+\alpha^2 \sum_{i,j}(2R_{ninj}^2 + (R_{1inj} + R_{ni1j})^2 + 2R_{1i1j}R_{ninj})$$

$$+2\alpha^2 \sum_i(R_{n1ni}^2 + R_{1n1i}^2) + \mathcal{C} + \mathcal{E}(\alpha,\beta),$$

and proceeding as before, it follows that

$$\sum_i (R^2_{n1ni} + R^2_{1n1i}) = 0.$$

Thus the constancy of sectional curvature follows from [79], [45].

Finally, note that the use of Euclidean rotations instead of hyperbolic rotations affects the signs in the coefficient of α^4 and hence the result above does not hold in the Riemannian case.

3.2 The Null Osserman Condition

In this section we define the Osserman condition for null vectors. In this case, the situation is more complicated than the nonnull one and a large number of examples can be constructed. We will follow an approach as discussed in the previous chapter. First of all, we are interested in obtaining the expression of the curvature tensors of null Osserman manifolds and then, by means of the second Bianchi identity, we will understand their geometries. Our main results correspond to those even-dimensional Lorentzian null Osserman manifolds.

Let (M, g) be a Lorentzian manifold and $u \in T_pM$ be a null vector. Then $u^\perp = (span\{u\})^\perp$ is a degenerate vector space of signature $(0 + \cdots +)$ containing $span\{u\}$. Hence the restriction of $R(\,\cdot\,, u)u$ to $u^\perp$ carries a useless information $R(u, u)u = 0$ together with difficulties of dealing with a degenerate vector space. But we can avoid these difficulties by quotienting out $span\{u\}$ from $u^\perp$. Now let $u \in T_pM$ be a null vector and let $\overline{u}^\perp = u^\perp/span\{u\}$ be the $(n-2)$-dimensional quotient space with the natural projection $\pi : u^\perp \to \overline{u}^\perp$. Also define an inner product on $\overline{u}^\perp$ by $\overline{g}(\overline{x}, \overline{y}) = g(x, y)$, where $x, y \in u^\perp$ with $\pi(x) = \overline{x}$, $\pi(y) = \overline{y}$. Note that $\overline{g}$ is a definite inner product on $\overline{u}^\perp$. Now we are ready to define the Jacobi operator with respect to a null vector.

Definition 3.2.1. Let (M, g) be a Lorentzian manifold and $u \in T_pM$ be a null vector. Then the linear map $\overline{R}_u : \overline{u}^\perp \to \overline{u}^\perp$ defined by

$$\overline{R}_u \overline{x} = \pi(R(x, u)u),$$

where $x \in u^\perp$ with $\pi(x) = \overline{x}$, is called the Jacobi operator with respect to u.

Here note that since $R(u, u)u = 0$, $\overline{R}_u$ is well-defined.

Proposition 3.2.1. Let (M, g) be a Lorentzian manifold and $u \in T_pM$ be a null vector. Then $\overline{R}_u$ is a self-adjoint linear map.

Proof. Let $x, y \in u^\perp$ with $\pi(x) = \overline{x}$, $\pi(y) = \overline{y}$. Then

$$\overline{g}(\overline{R}_u \overline{x}, \overline{y}) = \overline{g}(\pi(R(x, u)u), \pi(y)) = g(R(x, u)u, y)$$

$$= g(R(y, u)u, x) = \overline{g}(\pi(R(y, u)u), \pi(x))$$

$$= \overline{g}(\overline{R}_u \overline{y}, \overline{x}).$$

□

Remark 3.2.1. Note that, if (M, g) is a Lorentzian manifold and $u \in T_pM$ is a null vector then, since $\overline{R}_u$ is self-adjoint and $\overline{u}^\perp$ has definite inner product, $\overline{R}_u$ is diagonalizable.

In the definition of the Osserman condition (Definition 1.2.2), we used a canonically determined set of nonnull vectors on which the Osserman condition can be well-defined. But since $g(u, u) = 0$ for every null vector u, there is no canonical way of determining a set of null vectors on which the null Osserman condition can be defined similar to the Osserman condition. However, on a Lorentzian manifold, one can define a canonical set of null vectors by using a timelike vector since $g(u, z) \neq 0$ for every null vector u and timelike vector z at a point.

Definition 3.2.2. *Let (M, g) be a Lorentzian manifold and $z \in T_pM$ be a timelike unit vector. Then the null congruence $N(z)$ determined by z at p is defined by*

$$N(z) = \{u \in T_pM / g(u, u) = 0 \text{ and } g(u, z) = -1\}.$$

Now we can state the null Osserman condition with respect to a unit timelike vector.

Definition 3.2.3. *Let (M, g) be a Lorentzian manifold of $\dim M \geq 3$. Then (M, g) is called null Osserman with respect to a unit timelike $z \in T_pM$ if the characteristic polynomial of $\overline{R}_u$ is independent of $u \in N(z)$, that is, if the eigenvalues (counted with multiplicities) of $\overline{R}_u$ are independent of $u \in N(z)$.*

Next we state a theorem about rigidity of the null Osserman condition. Note that if (M, g) is a Lorentzian manifold which is of constant sectional curvature at $p \in M$ then since $\overline{R}_u = 0$, (M, g) is null Osserman with respect to every unit timelike $z \in T_pM$.

Proposition 3.2.2. *Let (M, g) be a Lorentzian manifold of $\dim M \geq 4$ and $z, z' \in T_pM$ be linearly independent timelike unit vectors. If (M, g) is null Osserman with respect to z and z' then (M, g) is of constant sectional curvature at $p \in M$.*

Proof. Let $u \in N(z)$. Then there exists $t \neq 0$ such that $tu \in N(z')$, and therefore $\overline{R}_{tu} = t^2 \overline{R}_u$. Hence if $\{c_i\}$ and $\{c_i'\}$ are the eigenvalues (counted with multiplicities) of $\overline{R}_u$ and $\overline{R}_{tu}$, respectively, then $c_i' = t^2 c_i$. But since z and z' are linearly independent, $N(z) \neq N(z')$ and it follows that $c_i' = c_i = 0$. Thus $\overline{R}_u = 0$ for every null $u \in T_pM$. Hence $\overline{g}(\overline{R}_u \overline{x}, \overline{x}) = g(R(x, u)u, x) = 0$ for every null $u \in T_pM$ and $x \in u^\perp$, where $\pi(x) = \overline{x}$. But then it follows that (M, g) is of constant sectional curvature at $p \in M$ by Lemma 3.1.1-(b). □

3.2.1 The Null Osserman Condition at a Point

Let (M, g) be a Lorentzian manifold and let $u \in T_pM$ be a null vector. Then a nondegenerate subspace $W \subset u^\perp$ (that is, the restriction of g to W is nondegenerate) with $dimW = dim\overline{u}^\perp$ is called a *geometric realization* of $\overline{u}^\perp$. Note that the restriction of g to W is a definite inner product and $\pi_{|w} : W \to \overline{u}^\perp$ is an isometry. Let $\overline{x} \in \overline{u}^\perp$ be an eigenvector of $\overline{R}_u$ with eigenvalue c. Then the vector $x \in W$ with $\pi(x) = \overline{x}$ is called the *geometrically realized* eigenvector of $\overline{R}_u$ in W corresponding to the eigenvalue c. Now by means of this geometric realization of $\overline{u}^\perp$, we define a bundle homomorphism by using $\overline{R}_u$ for all $u \in N(z)$.

Definition 3.2.4. *Let (M, g) be a Lorentzian manifold and $z \in T_pM$ be a timelike unit vector. Then the celestial sphere of z is defined by*

$$S(z) = \{x \in z^\perp / g(x, x) = 1\}.$$

Remark 3.2.2. Let $z \in T_pM$ be a timelike unit vector tangent to a Lorentzian manifold (M, g). Then (M, g) is called *spatially isotropic* [127] at $p \in M$ with respect to z if and only if for any given two vectors $x_1, x_2 \in (spann\{z\})^\perp$ with $g(x_1, x_1) = g(x_2, x_2)$, there is a local isometry ϕ such that

$$\phi(p) = p, \qquad \phi_{*_p}z = z, \qquad \phi_{*_p}x_1 = x_2.$$

Hence, if (M, g) is spatially isotropic at $p \in M$ with respect to $z \in T_pM$, then the group of local isometries preserving p and z acts transitively on the celestial sphere $S(z)$. Therefore, (M, g) is null Osserman at p with respect to $z \in T_pM$.

Let (M, g) be a Lorentzian manifold and $z \in T_pM$ be a timelike unit vector. Then note that for each $u \in N(z)$, there exists a unique $x \in S(z)$ such that $u = z + x$. Thus the map $\psi : N(z) \to S(z)$ defined by $\psi(u) = u - z = x$ is a diffeomorphism. Now let $u \in N(z)$. Then $W = u^\perp \cap z^\perp$ is the geometric realization of $\overline{u}^\perp$ contained in $z^\perp$. By identifying W with $T_xS(z)$ for each $u \in N(z)$, where $x = u - z$, we obtain an identification of $\overline{u}^\perp$ with $T_xS(z)$ for each $u \in N(z)$, where $x = u - z$. Hence the Jacobi operator $\overline{R}_u$ can be used to define a linear map $\mathcal{R}_x : T_xS(z) \to T_xS(z)$ so that $\mathcal{R}_xy \in T_xS(z)$ with $\pi(\mathcal{R}_xy) = \overline{R}_u\overline{y}$ via the above identification, where $x = u - z$ and $\pi(y) = \overline{y}$. Here also note that $\overline{R}_u$ and $\mathcal{R}_x$ above have the same characteristic polynomials. Furthermore $\mathcal{R}_x$ can be extended to a bundle homomorphism $\mathcal{R} : TS(z) \to TS(z)$ by $\mathcal{R}y = \mathcal{R}_xy$, where $y \in T_xS(z)$.

Theorem 3.2.1. *Let (M, g) be a Lorentzian manifold of $dimM = 2m \geq 4$. If (M, g) is null Osserman with respect to a timelike unit vector $z \in T_pM$ then $\overline{R}_u$ has only one eigenvalue, that is, $\overline{R}_u = c\overline{id}$ for all $u \in N(z)$.*

Proof. Let $c \in \mathbb{R}$ be an eigenvalue of $\overline{R}_u$ for all $u \in N(z)$. Then c is also an eigenfunction of $\mathcal{R} : TS(z) \to TS(z)$. Now consider the bundle homomorphism T defined on $S(z) \cong S^{2m-2}$ by $T = \mathcal{R} - cid$. Then by Theorem 2.1.2, $kerT = TS(z)$ and hence the eigenvalues of $\overline{R}_u$ are equal. $\square$

Theorem 3.2.2. *Let (M, g) be a Lorentzian manifold of $dimM = 4m + 3 \geq 7$. If (M, g) is null Osserman with respect to a timelike unit vector $z \in T_pM$ then either the Jacobi operators $\overline{R}_u$ have only one eigenvalue or they have exactly two distinct eigenvalues c_1 and c_2 with multiplicities 1 and $4m$, respectively, for all $u \in N(z)$.*

Proof. Let c_1 be an eigenvalue of $\overline{R}_u$ and, as in the proof of Theorem 3.2.1, consider the bundle homomorphism T_{c_1} defined on $S(z) \cong S^{4m+1}$ by $T_{c_1} = \mathcal{R} - c_1 id$. Then, by Theorem 2.1.3, $kerT_{c_1}$ is either a line subbundle or a hyper-subbundle of $TS(z)$ or $TS(z)$ itself. In the last two cases the claim follows immediately. Suppose $kerT_{c_1}$ is a line subbundle of $TS(z)$. Then there is another eigenvalue c_2 of $\overline{R}_u$. Consider the bundle homomorphism $T_{c_2} = \mathcal{R} - c_2 id$ on $TS(z)$. Again by Theorem 2.1.3, $kerT_{c_2}$ is either a line subbundle or a hyper-subbundle of $TS(z)$. But $kerT_{c_2}$ cannot be a line subbundle of $TS(z)$, since then $kerT_{c_1} \oplus kerT_{c_2}$ is a subbundle of $TS(z)$ of rank 2 in contradiction to Theorem 2.1.3. Thus $kerT_{c_2}$ is a hyper-subbundle of $TS(z)$ and hence c_2 is of multiplicity $4m$. $\square$

Now we study in more detail the different situations motivated by the above results. We begin with the following:

Definition 3.2.5. *Let (M, g) be a Lorentzian manifold of $dimM \geq 3$. A null vector $u \in T_pM$ is called isotropic if $\overline{R}_u = c_u \overline{id}$, where $c_u \in \mathbb{R}$. (M, g) is called null isotropic at $p \in M$ if each null $u \in T_pM$ is isotropic. (M, g) is called null isotropic if (M, g) is null isotropic at each $p \in M$.*

Proposition 3.2.3. *Let (M, g) be a 4-dimensional Lorentzian manifold. Then there exists an isotropic null vector at each $p \in M$.*

Proof. Let $z \in T_pM$ be a unit timelike vector and suppose that for each $u \in N(z)$, $\overline{R}_u$ has distinct eigenvalues $c_1(u)$ and $c_2(u)$. Then since $\mathcal{R}_x : T_xS(z) \to T_xS(z)$ has the same eigenvalues $c_1(x) = c_1(u)$ and $c_2(x) = c_2(u)$, where $x = u - z$, the roots of the characteristic polynomial of $\mathcal{R} : TS(z) \to TS(z)$, are distinct at each $x \in S(z)$. Hence they form smooth maps c_1 and c_2 on $S(z)$. Now if we set $T = \mathcal{R} - c_1 id$ to be a bundle homomorphism on $TS(z)$, then $kerT$ is a line subbundle of $S(z) \cong S^2$. But this is a contradiction to Theorem 2.1.2. $\square$

Remark 3.2.3. Let (M, g) be a Lorentzian manifold of dimension $n \geq 3$.

a) If $dimM = 3$ then (M, g) is null isotropic, since $dim\overline{u}^\perp = 1$ for all null $u \in T_pM$.

b) If $\overline{R}_u = c_u \overline{id}$ then $c_u = \frac{1}{n-2} Ric(u,u)$, where $u \in T_pM$ is a null vector. To show this, let $z, x \in T_pM$ be orthonormal timelike and spacelike vectors, respectively, with $u = z + x$. Also let $y_1, \ldots, y_{n-1} \in u^\perp$ be orthonormal spacelike vectors such that $\{z, x, y_1, \ldots, y_{n-2}\}$ is an orthonormal basis for T_pM. Then

$$(n-2)c_u = trace\overline{R}_u = \sum_{i=1}^{n-2} \overline{g}(\overline{R}_u \overline{y}_i, \overline{y}_i) = \sum_{i=1}^{n-2} g(R(y_i, u)u, y_i)$$

$$= -g(R(z,u)u, z) + g(R(x,u)u, x) + \sum_{i=1}^{n-2} g(R(y_i, u)u, y_i)$$

$$= Ric(u, u).$$

c) If $\overline{R}_u = c_u \overline{id}$ then $c_u = \frac{1}{g(x,x)} g(R(x,u)u, x)$ for every spacelike $x \in u^\perp$, where $u \in T_pM$ is a null vector. Conversely, if $c_u = \frac{1}{g(x,x)} g(R(x,u)u, x)$ for every spacelike $x \in u^\perp$ then $\overline{R}_u = c_u \overline{id}$. To show this last claim, note that the bilinear form $h(\overline{x}, \overline{y}) = g(\overline{R}_u \overline{x}, \overline{y})$ defined on $\overline{u}^\perp$ satisfies $h(\overline{x}, \overline{x}) = c_u$ for all $\overline{x} \in \overline{u}^\perp$. Thus by the polarization identity, $h = c_u \overline{g}$, and hence $\overline{R}_u = c_u \overline{id}$.

Next we determine the curvature tensor of a null isotropic Lorentzian manifold.

Lemma 3.2.1. *Let (M, g) be a semi-Riemannian manifold of index $0 < \nu < n$ and let F be an algebraic curvature map on T_pM. If $F(x, y, z, x) = 0$ for every nonnull orthonormal $x, y, z \in T_pM$ with $g(x, x) = -g(y, y) = -1$ then $F = \lambda R^0$, where $\lambda \in \mathbb{R}$. Note also the condition $g(x, x) = -g(y, y) = -1$ can be replaced by $g(x, x) = g(y, y) = 1$ or $g(x, x) = g(y, y) = -1$.*

Proof. See for example [101, p. 14], [45]. □

Now let (M, g) be a Lorentzian manifold of $dimM \geq 3$ and let $F^1 : T_pM \times T_pM \times T_pM \to T_pM$ be a trilinear map defined by

$$F^1(x, y)z = Ric(z, y)x - Ric(x, z)y$$

at each $p \in M$. Then it is easy to see that the map $R^1 : T_pM \times T_pM \times T_pM \times T_pM \to \mathbb{R}$ defined by

$$R^1(x, y, z, v) = \frac{1}{n-2}[g(F^1(x, y)z, v) + Ric(R^0(x, y)z, v)]$$

is an algebraic curvature map on T_pM. Also note that

$$R^1(x, u, u, x) = \frac{1}{n-2} Ric(u, u)g(x, x),$$

for every null $u \in T_pM$ and $x \in u^\perp$.

Remark 3.2.4. If (M, g) is a null isotropic Lorentzian manifold at $p \in M$ then

$$R^1(x, u, u, x) = c_u g(x, x)$$

for each null $u \in T_p M$ and $x \in u^\perp$, where $\overline{R}_u = c_u \overline{id}$.

Now we are ready to obtain the curvature characterization of a null isotropic Lorentzian manifold.

Theorem 3.2.3. [57] *Let (M, g) be a Lorentzian manifold of $\dim M \geq 3$. Then (M, g) is null isotropic at $p \in M$ if and only if*

$$R = R^1 - \frac{1}{(n-1)(n-2)} (Sc)_p R^0$$

where $(Sc)_p$ is the scalar curvature of (M, g) at $p \in M$.

Proof. "If" part of the proof can be directly shown from the expression of the curvature tensor above. To show the "only if" part of the proof, let $F = R - R^1$ be an algebraic curvature map on $T_p M$. Then $F(x, u, u, x) = 0$ for every null $u \in T_p M$ and $x \in u^\perp$ by Remarks 3.2.3-(c) and 3.2.4. Let $x \in T_p M$ be a unit spacelike vector and let $h(y, z) = F(x, y, z, x)$ be a bilinear form on $x^\perp$. Then $h(u, u) = 0$ for every null $u \in x^\perp$ and it follows from Lemma 1.2.1-(a) that $h(y, z) = \lambda_x g(y, z)$ on $x^\perp$, where $\lambda_x \in \mathbb{R}$. Thus $F(x, y, z, x) = \lambda_x g(y, z)$ on $x^\perp$ and hence $F(x, y, z, x) = 0$ for every orthogonal $y, z \in x^\perp$. Then by Lemma 3.2.1, it follows that $F = \lambda R^0$, that is, $R = R^1 + \lambda R^0$, where $\lambda \in \mathbb{R}$. Now it remains to determine the constant λ. For this, let $\{e_1, \ldots, e_n\}$ be an orthonormal basis for $T_p M$. Then

$$\sum_{i,j=1}^{n} g(e_i, e_i) g(e_j, e_j) R(e_i, e_j, e_j, e_i)$$

$$= \sum_{i,j=1}^{n} g(e_i, e_i) g(e_j, e_j)[R^1(e_i, e_j, e_j, e_i) + \lambda R^0(e_i, e_j, e_j, e_i)].$$

Hence $(Sc)_p = \frac{2(n-1)}{n-2}(Sc)_p + \lambda n(n-1)$ and it follows that $\lambda = -\frac{(Sc)_p}{(n-1)(n-2)}$. $\square$

Note that the Weyl tensor of a semi-Riemannian manifold of $\dim M \geq 3$ is defined to be the curvature tensor $\tilde{W}$ of

$$W = R - R^1 + \frac{1}{(n-1)(n-2)}(Sc)R^0. \tag{3.1}$$

Hence we also have the following characterization of the null isotropic Lorentzian manifolds:

Theorem 3.2.4. *Let (M, g) be a Lorentzian manifold of $dim M \geq 3$. Then, (M, g) is null isotropic at $p \in M$ if and only if $\tilde{W} = 0$ at $p \in M$.*

Proof. Immediate from Theorem 3.2.3. □

Corollary 3.2.1. *Let (M, g) be a Lorentzian manifold. If (M, g) is null isotropic at $p \in M$, then (M, g) is of constant sectional curvature at p if and only if (M, g) is Einstein at p.*

Proof. Immediate from Remark 3.2.3. □

The reader who is interested in the analysis of null isotropy in a more general setting of semi-Riemannian geometry is refered to [57].

Next we determine the curvature tensors of some Lorentzian null Osserman manifolds with respect to a timelike unit vector $z \in T_p M$ at $p \in M$. First we start with the simplest case of Lorentzian manifolds given by Theorem 3.2.1.

Theorem 3.2.5. *Let (M, g) be a Lorentzian manifold of $dim M \geq 3$. Then the following are equivalent:*

a) $\overline{R}_u = \mu \overline{id}$ *for every $u \in N(z)$, where $\mu \in \mathbb{R}$ and $z \in T_p M$ is a timelike unit vector, that is, (M, g) is null Osserman with respect to z and $\overline{R}_u$ has only one eigenvalue μ,*

b) $R(x, y)v = \lambda R^0(x, y)v + \mu R^0(\pi^\perp x, \pi^\perp y)\pi^\perp v$, *where $x, y, v \in T_p M$, $\lambda \in \mathbb{R}$ and $\pi^\perp : T_p M \to z^\perp$ is the orthogonal projection,*

c)

(c.1) $R(x, v)v = \lambda g(v, v)x$, *for every $v \in span\{z\}$ and $x \in z^\perp$,*

(c.2) $R(x, y)v = (\lambda + \mu)R^0(x, y)v$, *for every $x, y, v \in z^\perp$.*

Proof. $(a) \Rightarrow (b)$: Let $\overline{R}_u = \mu \overline{id}$ for every $u \in N(z)$ and let F be an algebraic curvature map on $T_p M$ defined by

$$F(x, y, v, w) = g(R(x, y)v, w) - \mu g(R^0(\pi^\perp x, \pi^\perp y)\pi^\perp v, \pi^\perp w),$$

where $x, y, v, w \in T_p M$. Now if $u = z + y \in N(z)$, where $y \in S(z)$, and x is a geometrically realized eigenvector of $\overline{R}_u$ in $u^\perp \cap z^\perp$, then $F(x, u, u, x) = 0$. Thus $F(x, u, u, x) = 0$ for every $x \in (span\{z, y\})^\perp$ and $u \in N(z)$. Hence it follows by Lemma 3.1.1-(b) that $F(x, y, v, w) = \lambda g(R^0(x, y)v, w)$. (Note here that the proof of Lemma 3.1.1 only depends on the curvature-like properties of $F(x, y, v, w) = g(R(x, y)v, w)$. Also see [101, p. 16] for a proof.) Thus the claim follows.

$(b) \Rightarrow (c)$: Straightforward computation.

$(c) \Rightarrow (a)$: Let $u = z + y \in N(z)$, where $y \in S(z)$. We show that every vector x in the geometric realization $u^\perp \cap z^\perp$ of $\overline{u}^\perp$ is a geometrically realized eigenvector of $\overline{R}_u$ corresponding to the eigenvalue μ. For this, first we have to show that $g(R(x, u)u, v) = 0$ for every $v \in (span\{z, x, y\})^\perp$. Note that since

$$g(R(x,u)u,v) = g(R(x,z)z,v) + g(R(x,z)y,v)$$

$$+g(R(x,y)z,v) + g(R(x,y)y,v) = 0,$$

x is a geometrically realized eigenvector of $\overline{R}_u$ in $u^\perp \cap z^\perp$. Also since $g(R(x,u)u,x) = -\lambda + \mu + \lambda = \mu$, x corresponds to the eigenvalue μ. Thus the claim follows. □

Although we are not able to completely describe the curvature of those null Osserman manifolds given by Theorem 3.2.1, one has the following:

Theorem 3.2.6. *Let (M,g) be a Lorentzian manifold of $\dim M = 2m + 1 \geq 5$. Let $z \in T_pM$ be a timelike unit vector and J be a complex structure on $z^\perp$ for which the induced inner product on $z^\perp$ is Hermitian. Then the following are equivalent:*

a)
  (a.1) (M,g) is null Osserman with respect to z and $\overline{R}_u$ has only two distinct eigenvalues c_1 and c_2 with multiplicities 1 and $2m - 2$, respectively,
  (a.2) If $u = z + y \in N(z)$, where $y \in S(z)$, then Jy is a geometrically realized eigenvector of $\overline{R}_u$ in $u^\perp \cap z^\perp$ corresponding to the eigenvalue c_1.
b) $R(x,y)v = \lambda R^0(x,y)v + \mu R^0(\pi^\perp x, \pi^\perp y)\pi^\perp v + \frac{\eta}{4}[R^0 - R^J](\pi^\perp x, \pi^\perp y)\pi^\perp v$, for every $x,y,v \in T_pM$, where $\mu = \frac{1}{3}(4c_2 - c_1)$, $\eta = \frac{3}{4}(c_1 - c_2)$, and $\pi^\perp : T_pM \to z^\perp$ is the orthogonal projection.
c)
  (c.1) $R(x,v)v = \lambda g(v,v)x$ for every $v \in span\{z\}$ and $x \in z^\perp$,
  (c.2) $R(x,y)v = (\lambda + \mu)R^0(x,y)v + \frac{\eta}{4}[R^0 - R^J](x,y)v$ for every $x,y,v \in z^\perp$, where $\mu + \eta = c_1$ and $\mu + \frac{1}{4}\eta = c_2$.

Proof. $(a) \Rightarrow (b)$: Let F be an algebraic curvature map on T_pM defined by

$$F(x,y,v,w) = g(R(x,y)v,w) - \mu g(R^0(\pi^\perp x, \pi^\perp y)\pi^\perp v, \pi^\perp w)$$

$$-\frac{\eta}{4}g([R^0 - R^J](\pi^\perp x, \pi^\perp y)\pi^\perp v, \pi^\perp w)$$

where $\mu = \frac{1}{3}(4c_2 - c_1)$ and $\eta = \frac{4}{3}(c_1 - c_2)$. Now if $u = z + y \in N(z)$, where $y \in N(z)$, and x is a geometrically realized eigenvector of $\overline{R}_u$ in $u^\perp \cap z^\perp$ corresponding to c_1 then $F(x,u,u,x) = c_1 - \mu - \eta = 0$. On the other hand, if x is a geometrically realized eigenvector of $\overline{R}_u$ in $u^\perp \cap z^\perp$ corresponding to c_2 then $F(x,u,u,x) = c_2 - \mu - \frac{1}{4}\eta = 0$. Also it can be shown similarly that if $x = \alpha x_1 + \beta x_2$, where x_1 and x_2 are geometrically realized eigenvectors of $\overline{R}_u$ in $u^\perp \cap z^\perp$ corresponding to c_1 and c_2, respectively, then $F(x,u,u,x) = 0$. Thus $F(x,u,u,x) = 0$ for every $x \in (span\{z,y\})^\perp$ and $u \in N(z)$. Hence it follows by Lemma 3.1.1-(b) that $F(x,y,v,w) = \lambda g(R^0(x,y)v,w)$. Thus the claim follows.

(b) $\Rightarrow$ (c): Straightforward computation.

(c) $\Rightarrow$ (a): Let $u = z + y \in N(z)$, where $y \in S(z)$ and $x = Jy$. We claim that x is a geometrically realized eigenvector of $\overline{R}_u$ in $u^\perp \cap z^\perp$ corresponding to the eigenvalue c_1. For this, it suffices to show that $g(R(x,u)u,v) = 0$ for every $v \in (span\{x,y,z\})^\perp$. Note that since

$$g(R(x,u)u,v) = g(R(x,z)z,v) + g(R(x,z)y,v)$$

$$+g(R(x,y)z,v) + g(R(x,y)y,v) = 0,$$

x is a geometrically realized eigenvector of $\overline{R}_u$ in $u^\perp \cap z^\perp$. Also since $g(R(x,u)u,x) = -\lambda + \mu + \lambda + \eta = \mu + \eta = c_1$, x corresponds to the eigenvalue c_1. On the other hand, if $x \in (span\{z,y,Jy\})^\perp$ then x is a geometrically realized eigenvector of $\overline{R}_u$ in $u^\perp \cap z^\perp$ corresponding to the eigenvalue c_2. To show this, it suffices to show that $g(R(x,u)u,Jy) = 0$ and $g(R(x,u)u,v) = 0$ for every $v \in (span\{x,y,Jy,z\})^\perp$. As above, $g(R(x,u)u,Jy) = 0$ and $g(R(x,u)u,v) = 0$, and hence x is a geometrically realized eigenvector of $\overline{R}_u$ in $u^\perp \cap z^\perp$. Also since $g(R(x,u)u,x) = -\lambda + \mu + \lambda + \frac{1}{4}\eta = \mu + \frac{1}{4}\eta = c_2$, x corresponds to the eigenvalue c_2. Thus the claim follows. $\square$

Proceeding in an analogous way as before, one has the following:

Theorem 3.2.7. *Let (M,g) be a Lorentzian manifold of $dim M = 4m+1 \geq 5$. Let $z \in T_pM$ be a unit timelike vector and $\phi = J_1, J_2, J_3$ be a quaternionic structure on $z^\perp$ for which the induced inner product on $z^\perp$ is Hermitian. Then the following are equivalent:*

a)
    (a.1) *(M,g) is null Osserman with respect to z and $\overline{R}_u$ has only two distinct eigenvalues c_1 and c_2 with multiplicities 3 and $4(m-1)$, respectively,*
    (a.2) *If $u = z + y \in N(z)$, where $y \in S(z)$, then ϕy is a geometrically realized eigenvector of $\overline{R}_u$ in $u^\perp \cap z^\perp$ corresponding to the eigenvalue c_1 for $\phi = J_1, J_2, J_3$,*

b) $R(x,y)v = \lambda R^0(x,y)v + \mu R^0(\pi^\perp x, \pi^\perp y)\pi^\perp v$

$$+\frac{\eta}{4}[R^0 - \sum_{i=1}^{3} R^{J_i}](\pi^\perp x, \pi^\perp y)\pi^\perp v,$$

for every $x,y,v \in T_pM$, where $\mu = \frac{1}{3}(4c_2 - c_1)$, $\eta = \frac{4}{3}(c_1 - c_2)$, and $\pi^\perp : T_pM \to z^\perp$ is the orthogonal projection,

c)
    (c.1) *$R(x,v)v = \lambda g(v,v)x$ for every $v \in span\{z\}$ and $x \in z^\perp$,*
    (c.2) *$R(x,y)v = (\lambda + \mu)R^0(x,y)v + \frac{\eta}{4}[R^0 - \sum_{i=1}^{3} R^{J_i}](x,y)v$ for every $x,y,v \in z^\perp$, where $\mu + \eta = c_1$ and $\mu + \frac{1}{4}\eta = c_2$.*

3.2.2 Decomposition Theorems

As noted before, many examples of Lorentzian null Osserman manifolds can be obtained by taking the Lorentzian product of $\mathbb{R}$ with a Riemannian point-wise Osserman manifold (N, g_N), that is, $(M, g) \cong (\mathbb{R} \times N, -dt^2 \oplus g_N)$. Then $Z = \partial/\partial t$ is the distinguished timelike unit vector field, with respect to which (M, g) is null Osserman at each point. Such a construction can easily be generalized to a *Lorentzian warped product*, where the warped product metric tensor becomes $-dt^2 \oplus fg_N$ and $f : \mathbb{R} \to \mathbb{R}$ is a strictly positive smooth function. (See [114] for more information on warped product metric tensors.)

Our purpose here is to show that Lorentzian warped product structure is indeed the local structure of many null Osserman Lorentzian manifolds. First of all, we extend the null Osserman condition to the whole manifold.

Definition 3.2.6. *Let (M, g) be a Lorentzian manifold of $\dim M \geq 3$ and let L be a (not necessarily continuous) timelike line subbundle of TM. Then (M, g) is called pointwise null Osserman with respect to L if (M, g) is null Osserman with respect to each timelike unit $z \in L$. Also (M, g) is called globally null Osserman with respect to L if (M, g) is pointwise null Osserman with respect to L and the characteristic polynomial of $\overline{R}_u$ is independent of unit $z \in L$.*

Proposition 3.2.4. *Let (M, g) be a Lorentzian null Osserman manifold with respect to $z \in T_pM$ at each $p \in M$. If (M, g) is not Einstein at each $p \in M$ then there exists a unique timelike smooth line subbundle L of TM such that (M, g) is pointwise null Osserman with respect to L.*

Proof. Let us also denote $z \in T_pM$ by z_p and define $L = \bigcup_{p \in M} (span\{z_p\})$.

First observe that, since (M, g) is not of constant sectional curvature at each $p \in M$, L is unique by Proposition 3.2.2. Also $Ric(u, u) = trace\overline{R}_u$ for each null $u \in TM$. (Cf. Remark 3.2.3-(b).) Now let $z \in L$ be a unit vector and, $u = z + x \in N(z)$ and $v = z - x \in N(z)$, where $x \in S(z)$. Then since (M, g) is null Osserman with respect to z, $Ric(u, u) = Ric(v, v)$, and since

$$Ric(u, u) = Ric(z, z) + 2Ric(z, x) + Ric(x, x),$$

$$Ric(v, v) = Ric(z, z) - 2Ric(z, x) + Ric(x, x),$$

it follows that $Ric(z, x) = 0$. Also since $L_p^\perp$ has definite induced inner product, the Ricci operator $\hat{Ric}$ is diagonalizable at p with an eigenvector z and all other eigenvectors correspond to the same eigenvalue are in $z^\perp$. That is at each $p \in M$, the Ricci tensor takes the form

$$Ric = \lambda(p)g_L \oplus \mu(p)g_{L^\perp},$$

where $\lambda(p)$ and $\mu(p)$ are the eigenvalues of $\hat{Ric}$ at $p \in M$, and g_L and $g_{L^\perp}$ are the restrictions of g to L and $L^\perp$, respectively. Next we show that the eigenvalues $\lambda(p)$ and $\mu(p)$ of the Ricci operator are distinct at each point $p \in M$. To show this, first note that since (M, g) is not Einstein at $p \in M$, $Ric(u, u) \neq 0$ for all $u \in N(z)$. (See Lemma 1.2.1-(a).) Hence if $u = z + x \in N(z)$, where $x \in S(z)$, then

$$0 \neq Ric(u, u) = Ric(z, z) + Ric(x, x) = \lambda g(z, z) + \mu g(x, x) = -\lambda + \mu$$

and hence $\lambda(p) \neq \mu(p)$ at each $p \in M$. Thus the eigenvalue $\lambda(p)$ corresponding to z_p is a simple root of the characteristic polynomial of $\hat{Ric}$ and therefore it is a smooth function on M. Then it follows that $L = ker(\hat{Ric} - \lambda id)$ is a smooth line subbundle of TM.   □

Remark 3.2.5. Note that, in Proposition 3.2.4, if $\dim M$ is even, then the assumption that (M, g) is not Einstein at each $p \in M$ can be weakened to that (M, g) is not of constant sectional curvature at each $p \in M$. Indeed, since $Ric(u, u) = (n - 2)c$ for every $u \in N(z)$, where $\overline{R}_u = c\bar{id}$, the proof of Proposition 3.2.4 remains valid if $c \neq 0$ at each $p \in M$, which is the case if (M, g) is not of constant sectional curvature at each $p \in M$.

Next, we give some local decomposition theorems for Lorentzian pointwise null Osserman manifolds.

Lemma 3.2.2. *Let (M, g) be a Lorentzian manifold of $\dim M \geq 4$. Let L be a timelike line subbundle of TM. If*

a) $R(x, y)v = \lambda R^0(x, y)v + \mu R^0(\pi^\perp x, \pi^\perp y)\pi^\perp v$ with $\mu \neq 0$ at each $p \in M$, where $\lambda, \mu \in C^\infty(M)$ (see Theorem 3.2.5),
b) $d\lambda|_{L^\perp} = 0$,

then (M, g) is locally a warped product $(I \times N, -dt^2 \oplus f g_N)$, where $I \subseteq \mathbb{R}$ is an open interval and (N, g_N) is a (Riemannian) real space form.

Proof. Let $Z \in \Gamma L$ be a local unit vector field and $X, Y, V \in \Gamma L^\perp$. Then, if $\perp_L$ denotes the component in L, we have

$$(\nabla_X R)(Y, Z)V = -d\lambda(X)g(Y, V)Z$$
$$-\mu[g(V, \nabla_X Z)Y - g(Y, V)\pi^\perp(\nabla_X Z)],$$

$$(\nabla_Y R)(Z, X)V = d\lambda(Y)g(X, V)Z$$
$$+\mu[g(V, \nabla_Y Z)X - g(X, V)\pi^\perp(\nabla_Y Z)],$$

$$(\nabla_Z R)(Y, X)V = d(\lambda + \mu)(Z)[g(Y, V)X - g(X, V)Y$$
$$+\mu[g(V, Y)(\nabla_Z X)^{\perp_L} - g(X, V)(\nabla_Z Y)^{\perp_L}].$$

Then in the second Bianchi identity

$$(\nabla_Z R)(X,Y)V + (\nabla_X R)(Y,Z)V + (\nabla_Y R)(Z,X)V = 0.$$

By taking $Y = V$ and $X \perp Y$, we obtain

$$0 = d(\lambda + \mu)(Z)g(Y,Y)X - d\lambda(X)g(Y,Y)Z$$
$$+\mu[g(Y,Y)(\nabla_Z X)^{\perp_L} + g(Y,Y)\pi^{\perp}(\nabla_X Z)$$
$$+g(Y,\nabla_Y Z)X - g(Y,\nabla_X Z)Y].$$

Taking the inner product of the above equation with Z, since $d\lambda_{|_{L^\perp}} = 0$ and $\mu \neq 0$ at each $p \in M$, we obtain

$$0 = d\lambda(X)g(Y,Y) + \mu g(Y,Y)g(\nabla_Z X, Z) = -\mu g(Y,Y)g(X, \nabla_Z Z)$$

and conclude that Z is a unit geodesic vector field.

Again, by taking the inner product of the above equation with X, we have that

$$0 = d(\lambda + \mu)(Z)g(Y,Y)g(X,X)$$
$$+g(Y,Y)g(\nabla_X Z, X) + g(X,X)g(Y, \nabla_X Z).$$

Hence

$$\frac{1}{\mu}d(\lambda + \mu)(Z) = -\frac{g(\nabla_X Z, X)}{g(X,X)} - \frac{g(\nabla_Y Z, Y)}{g(Y,Y)}.$$

But since $rank L^\perp \geq 3$, we conclude that

$$g(\nabla_X Z, X) = -\frac{1}{2\mu}d(\lambda + \mu)(Z)g(X,X).$$

Now by choosing $X \perp Y$, $X \perp V$, $Y \perp V$ in the second Bianchi identity and taking the inner product with Y, we obtain

$$0 = \mu g(\nabla_X Z, V)g(Y,Y).$$

But since $\mu \neq 0$ at each $p \in M$, it follows that $g(\nabla_X Z, V) = 0$. Thus we get

$$-\nabla Z = \frac{d(\lambda + \mu)(Z)}{2\mu}id$$

on $L^\perp$. Hence $L^\perp$ is integrable with totally umbilical integral manifolds. It also follows that the mean curvature vector field ω of these totally umbilical integral manifolds is normal parallel from the Codazzi equation since $(R(X,Y)V)^{\perp_L} = 0$ for $X,Y,V \in \Gamma L^\perp$. Then since the integral manifolds of L are totally geodesic and the integral manifolds of $L^\perp$ are totally umbilical with normal parallel mean curvature vector field too, (M,g) is locally a warped product $(I \times N, -dt^2 \oplus fg_N)$. (See for example, [121].) Also by the Gauss equations, if R_N is the curvature tensor of (N, g_N) then $R_N(x, y)v = R(x, y)v + g(\omega, \omega)R^0(x, y)v$ for $x, y, v \in TN$ and hence (N, g_N) is a (Riemannian) real space form with $R_N = (\lambda + \mu + g(\omega, \omega))R^0$.   □

Now we give a characterization of even dimensional Lorentzian globally null Osserman manifolds.

Lemma 3.2.3. *Let (M, g) be a Lorentzian pointwise null Osserman manifold with respect to a (not necessarily continuous) timelike line subbundle L of TM. If (M, g) is not of constant sectional curvature at each $p \in M$, then*

a) *L is a smooth line subbundle of TM.*
b) *$R(X, Y)V = \lambda R^0(X, Y)V + \mu R^0(\pi^\perp X, \pi^\perp Y)\pi^\perp V$, where $X, Y, V \in \Gamma TM$, $\lambda, \mu \in C^\infty(M)$, $\pi^\perp : TM \to L^\perp$ is the orthogonal projection,*
c) *$L^\perp$ is integrable,*
d) *$d(\lambda + \mu)|_{L^\perp} = 0$.*

Proof. It follows from Theorem 3.2.1 and Remark 3.2.5 that L is a smooth line subbundle of TM. Also by Theorems 3.2.1 and 3.2.5, $R(X, Y)V = \lambda R^0(X, Y)V + \mu R^0(\pi^\perp X, \pi^\perp Y)\pi^\perp V$ for every $X, Y, V \in \Gamma TM$ and, λ and μ are smooth functions on M. Thus these prove (a) and (b). To prove (c) and (d), let $X, Y, V, T \in \Gamma L^\perp$. Then, if $\pi : TM \to L$ is the orthogonal projection,

$$(\nabla_X R)(Y, V)T = d\lambda(X)[g(T, V)Y - g(T, Y)V]$$

$$+ d\mu(X)[g(T, V)Y - g(T, Y)V]$$

$$+ \mu[g(T, V)\pi(\nabla_X Y) - g(T, Y)\pi(\nabla_X V)].$$

Now by setting $T = V$, $Z \perp X$, $Z \perp Y$, we obtain from the second Bianchi identity $(\nabla_X R)(Y, V)T + (\nabla_Y R)(V, X)T + (\nabla_V R)(X, Y)T = 0$ that

$$\mu g(V, V)[\pi(\nabla_X Y) - \pi(\nabla_Y X)] = 0$$

and

$$g(V, V)[d(\lambda + \mu)(X)Y - d(\lambda + \mu)(Y)X] = 0$$

for every $X, Y, V \in \Gamma L^\perp$. Thus it follows that $\pi([X, Y]) = 0$ (since $\mu \neq 0$ at each $p \in M$) and $d(\lambda + \mu)|_{L^\perp} = 0$. These prove (c) and (d).   □

Hence, we have

Theorem 3.2.8. *Let (M, g) be a Lorentzian manifold of $\dim M = 2m \geq 4$. Let L be a (not necessarily continuous) timelike line subbundle of TM. If (M, g) is globally null Osserman with respect to L and is not of constant sectional curvature, then (M, g) is locally a warped product $(I \times N, -dt^2 \oplus f g_N)$, where $I \subseteq \mathbb{R}$ is an open interval and (N, g_N) is a (Riemannian) real space form.*

Proof. Immediate by Lemmas 3.2.2 and 3.2.3 since $d(\lambda + \mu)|_{L^\perp} = d\lambda|_{L^\perp}$. (Note that since (M, g) is globally null Osserman with respect to L, μ is constant on M.) □

Remark 3.2.6. Note that a Lorentzian warped product manifold as in the above theorem is not necessarily locally symmetric.

Before we state the next local decomposition theorem, first we fix our notation and emphasize the meaning of some assumptions to be made. Let (M, g) be a Lorentzian manifold of $dim M = 2m+1$ and let L be a timelike line subbundle of TM. Let $\pi^\perp : TM \to L^\perp$ be the orthogonal projection. Also let J be an *almost complex metric $L^\perp$-substructure*. That is, let J be a complex structure on $L^\perp$ for which the induced metric tensor $g_{L^\perp}$ is Hermitian. Define the *shape operator* $L_Z : L^\perp \to L^\perp$ of $L^\perp$ with respect to $Z \in \Gamma L$ by $L_Z = -\pi^\perp(\nabla Z)$, and define the *conjugate shape operator* $\overline{L}_Z : L^\perp \to L^\perp$ of $L^\perp$ with respect to $Z \in \Gamma L$ by $\overline{L}_Z = JL_Z$. Note that, if we define $J^\perp : TM \to L^\perp$ by $J^\perp = J \circ \pi^\perp$ then $\overline{L}_Z = -J^\perp(\nabla Z)$. Here it is worth remarking that if $L^\perp$ is integrable (or equivalently, L_Z is self-adjoint) then $\overline{L}_Z$ is skew-adjoint if and only if $JL_Z = L_Z J$ (or equivalently, L_Z is complex linear.) Moreover note that if $L^\perp$ is integrable and the integral manifolds are totally umbilical then $\overline{L}_Z$ is skew-adjoint.

Theorem 3.2.9. *Let (M, g) be a Lorentzian manifold of $dim M = 2m + 1 \geq 7$. Let L be a timelike line subbundle of TM and J be an almost complex metric $L^\perp$-substructure. If*

a) $R(x, y)v = \lambda R^0(x, y)v + \mu R^0(\pi^\perp x, \pi^\perp y)\pi^\perp v + \frac{\eta}{4}[R^0 - R^J](\pi^\perp x, \pi^\perp y)\pi^\perp v$, *with $\mu + \frac{1}{4}\eta \neq 0$ and $\eta \neq 0$ at each $p \in M$, where $\lambda, \mu, \eta \in C^\infty(M)$ (see Theorem 3.2.6),*

b) $d\lambda|_{L^\perp} = 0$,

c) $\overline{L}_Z$ *is skew-adjoint,*

then (M, g) is locally a warped product $(I \times N, -dt^2 \oplus fg_N)$, where $I \subseteq \mathbb{R}$ is an open interval and (N, J, g_N) is a (Riemannian) complex space form.

Proof. Let $Z \in \Gamma L$ be a local unit vector field and $X, Y, V \in \Gamma L^\perp$. Then, if $\perp_L$ denotes the component in L, we have

$$(\nabla_X R)(Y, Z)V = -d\lambda(X)g(Y, V)Z$$

$$-(\mu + \frac{1}{4}\eta)[g(V, \nabla_X Z)Y - g(Y, V)\pi^\perp(\nabla_X Z)]$$

$$-\frac{1}{4}\eta[g(J(\nabla_X Z), V)JY - g(JY, V)J^\perp(\nabla_X Z)$$

$$-2g(Y, J(\nabla_X Z))JV];$$

$$(\nabla_Y R)(Z, X)V = d\lambda(Y)g(V, X)Z$$

$$+(\mu + \frac{1}{4}\eta)[g(\nabla_Y Z, V)JX - g(JX, V)J^\perp(\nabla_Y Z)]$$

$$+\frac{1}{4}\eta[g(J(\nabla_Y Z), VV)JX - g(JX, V)J^\perp(\nabla_Y Z)$$

$$-2g(J(\nabla_Y Z), X)JV];$$

$$(\nabla_Z R)(X, Y)V = d(\lambda + \mu + \frac{1}{4}\eta)(Z)[g(V, Y)X - g(X, V)Y]$$

$$+\frac{1}{4}d\eta(Z)[g(JY, V)JX - g(JX, V)JY + 2g(X, JY)JV]$$

$$+(\mu + \frac{1}{4}\eta)[g(V, Y)(\nabla_Z X)^{\perp L} - g(V, X)(\nabla_Z Y)^{\perp L}]$$

$$+\frac{1}{4}\eta[g(\nabla_Z J^\perp)Y, V)JX + g(JY, V)(\nabla_Z J^\perp)X$$

$$-g((\nabla_Z J^\perp)X, V)JY - g(JX, V)(\nabla_Z J^\perp)Y$$

$$+2g(X, (\nabla_Z J^\perp)Y)JV + 2g(X, JY)(\nabla_Z J^\perp)V].$$

Then in the second Bianchi identity $(\nabla_Z R)(X, Y)V + (\nabla_X R)(Y, Z)V + (\nabla_Y R)(Z, X)V = 0$, by taking $Y = V$ and $X \perp Y$, $X \perp JY$, we obtain

$$0 = d(\lambda + \mu + \frac{1}{4}\eta)(Z)g(Y, Y)X - d\lambda(X)g(Y, Y)Z$$

$$+\mu[g(Y, Y)(\nabla_Z X)^{\perp L} + g(Y, Y)\pi^\perp(\nabla_X Z)$$

$$+g(Y, \nabla_Y Z)X - g(Y, \nabla_X Z)Y]$$

$$+\frac{1}{4}\eta[g(Y, Y)(\nabla_Z X)^{\perp L} + g(Y, Y)\pi^\perp(\nabla_X Z) + g(\nabla_Y Z, Y)X$$

$$-g(\nabla_X Z, Y)Y + g((\nabla_Z J^\perp)Y, Y)JX - g((\nabla_Z J^\perp)X, Y)JY$$

$$+g(J^\perp(\nabla_Z X), Y)JY - g(J^\perp(\nabla_X Z), Y)JY$$

$$+g(J^\perp(\nabla_Y Z), Y)JX + 2g(X, (\nabla_Z J^\perp)Y)JY$$

$$+2g(Y, J^\perp(\nabla_X Z))JY - 2g(\nabla_Y Z, JX)JY].$$

Taking the inner product of the above equation with Z, since $d\lambda_{|_{L^\perp}} = 0$ and $\mu + \frac{1}{4}\eta = c_2 \neq 0$ at each $p \in M$, we obtain

$$0 = d\lambda(X)g(Y, Y) + (\mu + \frac{1}{4}\eta)g(Y, Y)g(\nabla_Z X, Z)$$

$$= -(\mu + \frac{1}{4}\eta)g(Y, Y)g(X, \nabla_Z Z),$$

and we conclude that Z is a unit geodesic vector field.

Again by taking the inner product of the above equation with X, we have that

$$0 = d(\lambda + \mu + \frac{1}{4}\eta)(Z)g(Y,Y)g(X,X)$$

$$+ (\mu + \frac{1}{4}\eta)[g(Y,Y)g(\nabla_X Z, X) + g(X,X)g(\nabla_Y Z, Y)].$$

Hence

$$\frac{1}{\mu + \frac{1}{4}\eta}d(\lambda + \mu + \frac{1}{4}\eta)(Z) = -\frac{g(\nabla_X Z, X)}{g(X,X)} - \frac{g(\nabla_Y Z, Y)}{g(Y,Y)}.$$

But since $rank L^\perp \geq 6$, we conclude that

$$g(\nabla_X Z, X) = \frac{1}{2(\mu + \frac{1}{4}\eta)}d(\lambda + \mu + \frac{1}{4}\eta)(Z)g(X,X).$$

Now by choosing $X \perp Y$, $X \perp V$, $Y \perp V$, $X \perp JY$, $V \perp JY$ in the second Bianchi identity and taking the inner product by Y, we obtain

$$0 = \frac{1}{4}\eta g(JX, V)g((\nabla_Z J^\perp)Y, Y) + (\mu + \frac{1}{4}\eta)g(\nabla_X Z, V)g(Y,Y)$$

$$+ \frac{1}{4}\eta g(JX, V)g(J^\perp(\nabla_Y Z), Y).$$

But since $(\nabla_Z J^\perp)|_{L^\perp} : L^\perp \to L^\perp$ is skew-adjoint, $\overline{L}_Z$ is skew-adjoint and $\mu + \frac{1}{4}\eta \neq 0$ at each $p \in M$ by assumptions, it follows that $(\mu + \frac{1}{4}\eta)g(Y,Y)g(\nabla_X Z, V) = 0$ and hence $g(\nabla_X Z, V) = 0$. Thus we get

$$L_Z = -\nabla Z = \frac{d(\lambda + \mu + \frac{1}{4}\eta)(Z)}{2(\mu + \frac{1}{4}\eta)}id.$$

Thus $L^\perp$ is integrable with totally umbilical integral manifolds. Also it follows that the mean curvature vector field ω of these totally umbilical integral manifolds is normal parallel from the Codazzi equation since $(R(X,Y)V)^{\perp_L} = 0$ for $X,Y,V \in \Gamma L^\perp$. Then, since the integral manifolds of L are totally umbilical with normal parallel mean curvature vector field too, (M,g) is locally a warped product $(I \times N, -dt^2 \oplus fg_N)$. (See, for example, [121].) Also by the Gauss equations, if R_N is the curvature tensor of (N, g_N) then $R_N(x,y)v = R(x,y)v + g(\omega,\omega)R^0(x,y)v$ for $x,y,v \in TN$ and hence (N, g, J) is a generalized (Riemannian) complex space form with $R_N = (\lambda + \mu + \frac{1}{4}\eta + g(\omega,\omega))R^0 - \frac{1}{4}\eta R^J$, where $\eta \neq 0$. But a generalized complex space form of dimension ≥ 6 is necessarily a complex space form (cf. [140].) This completes the proof. □

Let (M,g) be a Lorentzian manifold of $dim M = 4m + 1$ and let L be a timelike line subbundle of TM. Let $\pi^\perp : TM \to L^\perp$ be the orthogonal projection. Also, let $\mathbb{Q}$ be an *almost quaternionic metric $L^\perp$-substructure* on $L^\perp$. That is, let $\mathbb{Q}$ be a quaternionic structure on $L^\perp$ for which the induced metric tensor $g_{L^\perp}$ on $L^\perp$ is Hermitian. (That is, $g_{L^\perp}(\phi X, \phi Y) = g_{L^\perp}(X, Y)$ for every $X, Y \in \Gamma L^\perp$ and $\phi \in \Gamma\mathbb{Q}$.) We can also define three conjugate shape operators L_Z^ϕ with respect to a local canonical basis $\phi = J_1, J_2, J_3$ for $\mathbb{Q}$ by $\overline{L}_Z^\phi = \phi L_Z$, where L_Z is the shape operator of $L^\perp$ with respect to $Z \in \Gamma L$. Also note that if we define $\phi^\perp : TM \to L^\perp$ by $\phi^\perp = \phi \circ \pi^\perp$ then $\overline{L}_Z^\phi = -\phi^\perp(\nabla Z)$.

Theorem 3.2.10. *Let (M, g) be a Lorentzian manifold of $\dim M = 4m + 1 \geq 13$. Let L be a timelike line subbundle of TM and let $\mathbb{Q}$ be an almost quaternionic metric $L^\perp$-substructure on $L^\perp$. If*

a) $R(x, y)v = \lambda R^0(x, y)v + \mu R^0(\pi^\perp x, \pi^\perp y)\pi^\perp v$

$$+ \frac{\eta}{4}[R^0 - \sum_{i=1}^{3} R^{J_i}](\pi^\perp x, \pi^\perp y)\pi^\perp v$$

with $\mu + \frac{1}{4}\eta \neq 0$ and $\eta \neq 0$ at each $p \in M$, where $\lambda, \mu, \eta \in C^\infty(M)$ (see Theorem 3.2.7),

b) $d\lambda|_{L^\perp} = 0$,

c) $\overline{L}_Z^\phi$ *is skew-symmetric, where $\phi = J_1, J_2, J_3$ is a canonical local basis for $\mathbb{Q}$,*

then (M, g) is locally a warped product $(I \times N, -dt^2 \oplus fg_N)$, where $I \subseteq \mathbb{R}$ is an open interval and $(N, \mathbb{Q}, g_N)$ is a (Riemannian) quaternionic space form.

Proof. As in the proof of Theorem 3.2.9, we use the second Bianchi identity. At the final step, we use the fact that a generalized quaternionic space form is a quaternionic space form if its dimension is ≥ 12 (cf. [78].) $\qquad\square$

Remark 3.2.7. Note here that there are also indecomposable examples of null Osserman Lorentzian manifolds. To this end, we first recall the notion of a contact structure. A *contact manifold* is a $(2m + 1)$-dimensional manifold M equipped with a global 1-form η such that $\eta \wedge (d\eta)^m \neq 0$ everywhere on M, where $(d\eta)^m = d\eta \wedge \ldots \wedge d\eta$ (m-times.) Also such a manifold has an underlying *almost contact structure* (φ, ξ, η), where ξ is a global vector field on M (called the characteristic vector field) and φ is a $(1, 1)$-tensor field such that

$$\eta(\xi) = 1, \quad \varphi\xi = 0, \quad \eta\varphi = 0, \quad \varphi^2 = -id + \eta \otimes \xi.$$

A semi-Riemannian metric g is called *adapted* to the almost contact structure if $g(\varphi X, \varphi Y) = g(X, Y) - \eta(X)\eta(Y)$ for every $X, Y \in \Gamma TM$ (see [13],[28] for more information on almost contact structures.) If M is equipped with a Lorentzian metric tensor such that the characteristic vector field is timelike, then $(M, \varphi, \xi, \eta, g)$ is called a *Lorentzian almost contact manifold* [51]. A *K-contact* manifold is a contact manifold with an adapted metric tensor g such that ξ is a Killing vector. Moreover, if the almost complex structure J

on $M \times \mathbb{R}$ defined by $J(X, a\frac{\partial}{\partial t}) = (\varphi X - a\xi, \eta(X)\frac{\partial}{\partial t})$ is integrable, then it is called an *indefinite Sasakian manifold* [138]. Therefore, on any indefinite Sasakian manifold, the tensor field φ may be interpreted as an almost complex structure on the transverse of the foliation determined by the characteristic vector field. Moreover, note that the contactness condition means that the transverse of the characteristic foliation is never integrable, and thus, M does not split as a (warped)-product as discussed above.

A plane P tangent to a Lorentzian Sasakian manifold is called a φ-plane if it remains invariant under the action of φ and is orthogonal to ξ. Then the φ-sectional curvature is defined to be the restriction of the sectional curvature to φ-planes. A Lorentzian Sasakian manifold is called a *Lorentzian Sasakian space form* if φ-*sectional curvature* is constant. The curvature tensor of a Lorentzian Sasakian space form is given by

$$R(X, Y, Z, W) = -[g(Y, Z)g(X, W) - g(X, Z)g(Y, W)]$$
$$+ \frac{c+1}{4}[g(Y, Z)g(X, W) - g(X, Z)g(Y, W) + g(\varphi Y, Z)g(\varphi X, W)$$
$$-g(\varphi X, Z)g(\varphi Y, W) + 2g(X, \varphi Z, W) - \eta(Y)\eta(Z)g(X, W)$$
$$-\eta(X)\eta(W)g(Y, Z) + \eta(X)\eta(Z)g(Y, W) + \eta(Y)\eta(W)g(X, Z)],$$

where c is the constant value of the φ-sectional curvature. Now, a straightfoward calculation shows that Lorentzian Sasakian space forms are globally null Osserman with respect to the characteristic vector field.

Finally, note that, locally any Sasakian manifold is a line bundle over a Kähler manifold. Therefore, Lorentzian Sasakian manifolds are timelike line bundles over indefinite Kähler manifolds and it turns out that, a Lorentzian Sasakian manifold is globally null Osserman with respect to the characteristic vector field if and only if the base manifold of the fibration is a Kähler Osserman manifold.

4. Four-Dimensional Semi-Riemannian Osserman Manifolds with Metric Tensors of Signature (2, 2)

Since any semi-Riemannian Osserman manifold is Einstein and 3-dimensional Einstein manifolds are real space forms, 4-dimensional semi-Riemannian case is of special interest since it is the first non-trivial dimension to be considered. The simplest generic signature of a semi-Riemannian metric tensor is $(2, 2)$ on a 4-dimensional manifold. Such a metric tensor presents certain similarities to Riemannian metric tensors, which is a consequence of the fact that the Hodge star operator $*$ is an involution of the bundle of two-forms $\bigwedge^2$ on the manifold, contrary to the Lorentzian signature where $*^2 = -id_{\bigwedge^2}$. However, they exhibit notable differences from the Riemannian case. For instance, the existence of a metric tensor of signature $(2, 2)$ on compact manifolds is a rather restrictive condition (see [105], [106].)

The purpose of this chapter is to understand the consequences of the Osserman condition in the case of a metric tensor of signature $(2, 2)$ on a 4-dimensional manifold. Note that the examples showing the existence of non-symmetric Osserman manifolds to be constructed in Section 4.1 indicates the essential difference of this signature from both Riemannian and Lorentzian cases previously investigated. The fact that a self-adjoint linear map of an indefinite inner product space is not necessarily diagonalizable suggests that we pay attention to the minimal polynomials of the Jacobi operators. This leads to the consideration of the so-called Jordan-Osserman manifolds. Thus we devote Section 4.2 to a study of Jordan-Osserman manifolds with metric tensors of signature $(2, 2)$. As in the Riemannian case, we distinguish the algebraic, pointwise and global Osserman conditions. In doing that, we will follow the approach of Blažić, Bokan and Rakić [18], who obtained a complete description of the Osserman algebraic curvature tensors in signature $(2, 2)$ and most of the results on the local characterization of $(2, 2)$-Osserman manifolds to be presented in this chapter.

Note here that 4-dimensional semi-Riemannian Osserman manifolds are not yet completely classified. However, they are well understood if the manifold is further assumed to be locally symmetric.

4.1 Nonsymmetric Semi-Riemannian Osserman Manifolds with Metric Tensors of Signature $(2,2)$

To this point, the only known examples of semi-Riemannian Osserman manifolds are locally symmetric. However, as we will show soon, there are many nonsymmetric semi-Riemannian Osserman manifolds not even locally homogeneous when the metric tensor is allowed to possess generic signatures.

In this section, we construct a family of 4-dimensional semi-Riemannian Osserman manifolds with metric tensors of signature $(2,2)$. In all the cases given below, the characteristic polynomial of the Jacobi operators is $p_\lambda(R_x)$ $= \lambda^4$. However the behavior of the minimal polynomials differs and varies from $m_\lambda(R_x) = \lambda$ to $m_\lambda(R_x) = \lambda^3$. Such examples are constructed on $\mathbb{R}^4$ by considering special kind of semi-Riemannian metric tensors (see (4.1) below.) Before describing them in detail, let us give some motivation for their construction.

As already known, any semi-Riemannian Osserman metric tensor is necessarily Einstein (cf. Remark 1.3.1.) The simplest examples of Einstein manifolds are those hypersurfaces in real space forms. Such hypersurfaces were studied by Magid [103], who showed that their shape operators are either diagonalizable, or a complex structure on the hypersurface can be defined (after rescaling) or they are nilpotent. The study of this last kind of hypersurfaces suggests the following family of metric tensors of signature $(2,2)$.

Let $M = \mathbb{R}^4$ with usual coordinates (x_1, x_2, x_3, x_4). Then

$$
\begin{aligned}
g_{(f_1,f_2)} = {} & x_3 f_1(x_1, x_2) dx_1 \otimes dx_1 + x_4 f_2(x_1, x_2) dx_2 \otimes dx_2 \\
& + a[dx_1 \otimes dx_2 + dx_2 \otimes dx_1] \\
& + b[dx_1 \otimes dx_3 + dx_3 \otimes dx_1 + dx_2 \otimes dx_4 + dx_4 \otimes dx_2]
\end{aligned}
\tag{4.1}
$$

defines a semi-Riemannian metric tensor on $\mathbb{R}^4$ of signature $(2,2)$ for all real constants a and $b \neq 0$ and for all smooth real valued functions f_1, f_2.

Further assume that f_1 and f_2 satisfy

$$
\frac{\partial f_1}{\partial x_2} + \frac{\partial f_2}{\partial x_1} = 0.
\tag{4.2}
$$

Then the Christoffel symbols

$$
\Gamma_{ij}^k = \frac{1}{2} \sum g_{(f_1,f_2)}^{kl} \left\{ \frac{\partial g_{(f_1,f_2)_{il}}}{\partial x_j} + \frac{\partial g_{(f_1,f_2)_{jl}}}{\partial x_i} - \frac{\partial g_{(f_1,f_2)_{ij}}}{\partial x_l} \right\}
$$

of $g_{(f_1,f_2)}$ are given by

$$\Gamma_{11}^1 = -\frac{1}{2b}f_1, \qquad \Gamma_{11}^3 = \frac{1}{2b}x_3\frac{\partial f_1}{\partial x_1} + \frac{1}{2b^2}x_3 f_1^2,$$

$$\Gamma_{12}^3 = \frac{1}{2b}x_3\frac{\partial f_1}{\partial x_2}, \qquad \Gamma_{11}^4 = -\frac{1}{2b}x_3\frac{\partial f_1}{\partial x_2} + \frac{a}{2b^2}f_1,$$

$$\Gamma_{12}^4 = -\frac{1}{2b}x_4\frac{\partial f_1}{\partial x_2}, \qquad \Gamma_{13}^3 = \frac{1}{2b}f_1,$$

$$\Gamma_{22}^2 = -\frac{1}{2b}f_2, \qquad \Gamma_{22}^3 = \frac{1}{2b}x_4\frac{\partial f_1}{\partial x_2} + \frac{a}{2b^2}f_2,$$

$$\Gamma_{24}^4 = \frac{1}{2b}f_2, \qquad \Gamma_{22}^4 = \frac{1}{2b}x_4\frac{\partial f_2}{\partial x_2} + \frac{1}{2b^2}x_4 f_2^2,$$

and the others are zero. Hence, if we put $E_i = \frac{\partial}{\partial x_i}$, $i = 1, 2, 3, 4$, then the only nonvanishing covariant derivatives are given by

$$\nabla_{E_1} E_1 = -\frac{1}{2b}f_1 E_1 + \left[\frac{1}{2b}x_3\frac{\partial f_1}{\partial x_1} + \frac{1}{2b^2}x_3 f_1^2\right] E_3$$

$$+ \left[-\frac{1}{2b}x_3\frac{\partial f_1}{\partial x_2} + \frac{a}{2b^2}f_1\right] E_4,$$

$$\nabla_{E_1} E_2 = \frac{1}{2b}x_3\frac{\partial f_1}{\partial x_2} E_3 - \frac{1}{2b}x_4\frac{\partial f_1}{\partial x_2} E_4,$$

$$\nabla_{E_1} E_3 = \frac{1}{2b}f_1 E_3, \qquad\qquad (4.3)$$

$$\nabla_{E_2} E_2 = -\frac{1}{2b}f_2 E_2 + \left[\frac{1}{2b}x_4\frac{\partial f_1}{\partial x_2} + \frac{a}{2b^2}f_2\right] E_3$$

$$+ \left[\frac{1}{2b}x_4\frac{\partial f_2}{\partial x_2} + \frac{1}{2b^2}x_4 f_2^2\right] E_4,$$

$$\nabla_{E_2} E_4 = \frac{1}{2b}f_2 E_4.$$

From (4.3) it follows that the only nonvanishing components of the curvature tensor of $(\mathbb{R}^4, g_{(f_1, f_2)})$ are given by

$$R(E_1, E_2)E_3 = -\frac{1}{2b}\frac{\partial f_1}{\partial x_2}E_3, \qquad R(E_1, E_2)E_4 = -\frac{1}{2b}\frac{\partial f_1}{\partial x_2}E_4,$$

$$R(E_1, E_3)E_1 = \frac{1}{2b}\frac{\partial f_1}{\partial x_2}E_4, \qquad R(E_1, E_3)E_2 = -\frac{1}{2b}\frac{\partial f_1}{\partial x_2}E_3,$$

$$R(E_2, E_4)E_1 = \frac{1}{2b}\frac{\partial f_1}{\partial x_2}E_4, \qquad R(E_2, E_4)E_2 = -\frac{1}{2b}\frac{\partial f_1}{\partial x_2}E_3,$$

$$R(E_1, E_2)E_1 = \frac{1}{2b}\frac{\partial f_1}{\partial x_2}E_1 - \frac{1}{2b^2}x_3 f_1 \frac{\partial f_1}{\partial x_2}E_3$$

$$+\frac{1}{4b^3}\left[2b^2 x_3 \frac{\partial^2 f_1}{\partial x_2^2} - 2b^2 x_4 \frac{\partial^2 f_1}{\partial x_1 \partial x_2}\right. \tag{4.4}$$

$$\left.+b(x_3 f_2 - x_4 f_1 - 2a)\frac{\partial f_1}{\partial x_2} - a f_1 f_2\right] E_4,$$

$$R(E_1, E_2)E_2 = \frac{1}{2b}\frac{\partial f_1}{\partial x_2}E_2 - \frac{1}{2b^2}x_4 f_2 \frac{\partial f_1}{\partial x_2}E_4$$

$$-\frac{1}{4b^3}\left[2b^2 x_3 \frac{\partial^2 f_1}{\partial x_2^2} - 2b^2 x_4 \frac{\partial^2 f_1}{\partial x_1 \partial x_2}\right.$$

$$\left.+b(x_3 f_2 - x_4 f_1 + 2a)\frac{\partial f_1}{\partial x_2} - a f_1 f_2\right] E_3.$$

Now, if $X = \sum_{i=1}^{4}\alpha_i E_i$ is a vector field on $\mathbb{R}^4$, then the associated linear operator $R_X = R(\cdot, X)X$ defines a self-adjoint bundle homomorphism on $T\mathbb{R}^4$. Moreover, the matrix associated to R_X with respect to the basis $\{E_1, E_2, E_3, E_4\}$ is given by

$$R_X = \begin{pmatrix} A & 0 \\ B & {}^t A \end{pmatrix}, \tag{4.5}$$

where

$$A = \frac{1}{2b}(\frac{\partial f_1}{\partial x_2})\begin{pmatrix} \alpha_1\alpha_2 & -\alpha_1^2 \\ \alpha_2^2 & -\alpha_1\alpha_2 \end{pmatrix} \tag{4.6}$$

and the entries of B are given by

$$b_{11} = \frac{1}{4b^3}\left[-2b^2\alpha_2^2 x_3 \frac{\partial^2 f_1}{\partial x_2^2} + 2b^2\alpha_2^2 x_4 \frac{\partial^2 f_1}{\partial x_1 \partial x_2}\right.$$

$$-b(2\alpha_1\alpha_2 x_3 f_1 + \alpha_2^2 x_3 f_2 - \alpha_2^2 x_4 f_1 + 4b\alpha_2\alpha_3 + 2a\alpha_2^2)\frac{\partial f_1}{\partial x_2}$$

$$\left.+a\alpha_2^2 f_1 f_2\right],$$

$$b_{12} = \frac{1}{4b^3}\left[2b^2\alpha_1\alpha_2 x_3 \frac{\partial^2 f_1}{\partial x_2^2} - 2b^2\alpha_1\alpha_2 x_4 \frac{\partial^2 f_1}{\partial x_1 \partial x_2}\right.$$

$$+b(2\alpha_1^2 x_3 f_1 + \alpha_1\alpha_2 x_3 f_2 - \alpha_1\alpha_2 x_4 f_1 + 2b\alpha_1\alpha_3 + 2a\alpha_1\alpha_2$$

$$\left.-2b\alpha_2\alpha_4)\frac{\partial f_1}{\partial x_2} - a\alpha_1\alpha_2 f_1 f_2\right],$$

$$b_{21} = \frac{1}{4b^3}\left[2b^2\alpha_1\alpha_2 x_3 \frac{\partial^2 f_1}{\partial x_2^2} - 2b^2\alpha_1\alpha_2 x_4 \frac{\partial^2 f_1}{\partial x_1 \partial x_2}\right.$$

$$-b(\alpha_1\alpha_2 x_4 f_1 + 2\alpha_2^2 x_4 f_2 - \alpha_1\alpha_2 x_3 f_2 - 2b\alpha_1\alpha_3 + 2a\alpha_1\alpha_2$$

$$\left.+2b\alpha_2\alpha_4)\frac{\partial f_1}{\partial x_2} - a\alpha_1\alpha_2 f_1 f_2\right],$$

$$b_{22} = \frac{1}{4b^3}\left[-2b^2\alpha_1^2 x_3 \frac{\partial^2 f_1}{\partial x_2^2} + 2b^2\alpha_1^2 x_4 \frac{\partial^2 f_1}{\partial x_1 \partial x_2}\right.$$

$$+b(2\alpha_1\alpha_2 x_4 f_2 + \alpha_1^2 x_4 f_1 - \alpha_1^2 x_3 f_2 + 4b\alpha_1\alpha_4 + 2a\alpha_1^2)\frac{\partial f_1}{\partial x_2}$$

$$\left.+a\alpha_1^2 f_1 f_2\right].$$

(4.7)

Hence we have the following:

Theorem 4.1.1. $(\mathbb{R}^4, g_{(f_1,f_2)})$ *is a semi-Riemannian Osserman manifold with metric tensor of signature* $(2,2)$. *Moreover the characteristic polynomial of the operators* R_X *is* $p_\lambda(R_X) = \lambda^4$ *and the minimal polynomial* $m_\lambda(R_X)$ *of the operators* R_X *satisfies*

(i) $m_\lambda(R_X) = \lambda^3$ *at the points where* $\dfrac{\partial f_1}{\partial x_2} \neq 0$.

(ii) *At the points where* $\dfrac{\partial f_1}{\partial x_2} = 0$, *the function*

$$F(x_1, x_2, x_3, x_4) = 2b^2 \frac{\partial}{\partial x_2}(x_3 \frac{\partial f_1}{\partial x_2} - x_4 \frac{\partial f_1}{\partial x_1}) - af_1 f_2 \qquad (4.8)$$

determines the minimal polynomial as follows:

(ii.a) $(\mathbb{R}^4, g_{(f_1,f_2)})$ *has zero curvature at the points where* $\dfrac{\partial f_1}{\partial x_2}$ *and* F *vanish,* $(m_\lambda(R_X) = \lambda)$, (ii.b) *the minimal polynomial is* $m_\lambda(R_X) = \lambda^2$ *at the points where* $\dfrac{\partial f_1}{\partial x_2} = 0$ *and F is different from zero.*

Proof. It follows from the expression (4.5) of R_X, where X is a nonnull vector field on $\mathbb{R}^4$, and (4.6) that its characteristic polynomial satisfies $p_\lambda(R_X) = \det(R_X - \lambda I_4) = \lambda^4$. Thus all eigenvalues are zero. This proves that $(\mathbb{R}^4, g_{(f_1,f_2)})$ is globally Osserman.

Moreover, by straightforward calculations using (4.6) and (4.7), one obtains

$$R_X^2 = \begin{pmatrix} 0 & 0 \\ BA + {}^t\!AB & 0 \end{pmatrix} = \frac{1}{4b^3} g(X,X) \left(\frac{\partial f_1}{\partial x_2}\right)^2 \begin{pmatrix} 0 & 0 & 0 & 0 \\ 0 & 0 & 0 & 0 \\ -\alpha_2^2 & \alpha_1\alpha_2 & 0 & 0 \\ \alpha_1\alpha_2 & -\alpha_1^2 & 0 & 0 \end{pmatrix}$$

and

$$R_X^3 = \begin{pmatrix} 0 & 0 \\ (BA + {}^t\!AB)A & 0 \end{pmatrix} = 0$$

for all vector fields X on $\mathbb{R}^4$. Thus the minimal polynomial becomes $m_\lambda(R_X) = \lambda^3$ whenever $\frac{\partial f_1}{\partial x_2} \neq 0$. At the points where $\frac{\partial f_1}{\partial x_2} = 0$, the minimal polynomial is $m_\lambda(R_X) = \lambda^2$ or $m_\lambda(R_X) = \lambda$ depending on whether the function F above vanishes, since in this case (4.5) becomes

$$R_X = \frac{1}{4b^3} F \begin{pmatrix} 0 & 0 & 0 & 0 \\ 0 & 0 & 0 & 0 \\ -\alpha_2^2 & \alpha_1\alpha_2 & 0 & 0 \\ \alpha_1\alpha_2 & -\alpha_1^2 & 0 & 0 \end{pmatrix}.$$

□

Remark 4.1.1. Condition (4.2) imposed on the functions f_1 and f_2 allows us to simplify the calculations above. Proceeding in the same way as before, it can be shown that such a condition is indeed equivalent to $(M, g_{(f_1,f_2)})$ being Osserman.

Theorem 4.1.2. $(\mathbb{R}^4, g_{(f_1,f_2)})$ *is a locally symmetric space if and only if, in addition to condition (4.2), the functions f_1 and f_2 are solutions of the following partial differential equations on $\mathbb{R}^4$:*

(i) $\dfrac{\partial^2 f_1}{\partial x_1 \partial x_2} + \dfrac{1}{2b} f_1 \dfrac{\partial f_1}{\partial x_2} = 0,$

(ii) $\dfrac{\partial^2 f_1}{\partial x_2^2} + \dfrac{1}{2b} f_2 \dfrac{\partial f_1}{\partial x_2} = 0,$

(iii) $\dfrac{1}{4b} a \left[3f_1 \dfrac{\partial f_1}{\partial x_2} - f_2 \dfrac{\partial f_1}{\partial x_1} - \dfrac{1}{b} f_1^2 f_2 \right] - x_3 \left(\dfrac{\partial f_1}{\partial x_2} \right)^2 = 0,$

(iv) $\dfrac{1}{4b} a \left[3f_2 \dfrac{\partial f_1}{\partial x_2} + f_1 \dfrac{\partial f_2}{\partial x_2} + \dfrac{1}{b} f_1 f_2^2 \right] + x_4 \left(\dfrac{\partial f_1}{\partial x_2} \right)^2 = 0.$

Proof. It follows from (4.3) and (4.4) after a long but straightforward calculation.   □

Remark 4.1.2. Using the metric tensors described above (4.1), one can construct examples of nonsymmetric semi-Riemannian globally Osserman manifolds of signature $(2,2)$, where the behavior of the minimal and characteristic polynomials of the Jacobi operators is as follows:

1. The minimal polynomial is $m_\lambda(R_X) = \lambda^3$ by choosing $f_1(x_1, x_2) = x_2$, $f_2(x_1, x_2) = -x_1$ and for any value of a with $b \neq 0$.

2. The minimal polynomial is $m_\lambda(R_X) = \lambda^2$ by choosing $f_1(x_1, x_2) = k_1$, $f_2(x_1, x_2) = k_2$ and for any value of a and b, and for all the constants k_1, k_2, a, b different from zero.

3. If we take $f_1(x_1, x_2) = x_1$, $f_2(x_1, x_2) = k$ and a, b and k are different from zero, the operators R_X of the resulting manifold have minimal polynomial $m_\lambda(R_X) = \lambda^2$ at any point with $x_1 \neq 0$ and $m_\lambda(R_X) = \lambda$ at the points where $x_1 = 0$.

4. For the special choice of $f_1(x_1, x_2) = x_1 x_2$, $f_2(x_1, x_2) = -\frac{1}{2} x_1^2$ and for any constants a, b with $b \neq 0$, the minimal polynomial corresponding to the operators R_X of the resulting manifold is $m_\lambda(R_X) = \lambda^3$ at any point with $x_1 \neq 0$ and $m_\lambda(R_X) = \lambda^2$ at the points where $x_1 = 0$, $x_4 \neq 0$ and $m_\lambda(R_X) = \lambda$ at points where $x_1 = x_4 = 0$.
   Note that, on the open subset determined by $x_4 \neq 0$, the minimal polynomial of the operators R_X varies from $m_\lambda(R_X) = \lambda^3$ to $m_\lambda(R_X) = \lambda^2$ according to x_1 being different from or equal to zero.

5. Define $f_1(x_1, x_2) = x_1^3 x_2^3$ and $f_2(x_1, x_2) = -\frac{3}{4} x_1^4 x_2^2$. Then, for any value of a and $b \neq 0$, the operators R_X of $(\mathbb{R}^4, g_{(f_1, f_2)})$ have minimal polynomial $m_\lambda(R_X) = \lambda^3$ at the points with $x_1 x_2 \neq 0$ and $m_\lambda(R_X) = \lambda$ at all the points with $x_1 x_2 = 0$.

Remark 4.1.3. The previous examples show that, in general, the roots of the minimal polynomial and their multiplicities may change from point to point. However they are necessarily constant at each point since conditions given in Theorem 4.1.1 do not depend on any particular direction.

Remark 4.1.4. Let N be the open subset of $\mathbb{R}^4$ determined by $x_1 \neq 0$ and x_2 $\neq 0$ equipped with the metric tensor (4.1) given by the functions $f_1(x_1, x_2)$ $= \frac{b}{x_1}$ and $f_2(x_1, x_2) = \frac{b}{x_2}$, $a \neq 0$. Then $(N, g_{(f_1, f_2)})$ is a 4-dimensional semi-Riemannian globally Osserman manifold with $p_\lambda(R_X) = \lambda^4$ and $m_\lambda(R_X)$ $= \lambda^2$ by the direct application of Theorem 4.1.1. Moreover, it follows from Theorem 4.1.2 that $(N, g_{(f_1, f_2)})$ is locally symmetric.

4.2 Semi-Riemannian Osserman and Jordan-Osserman Manifolds with Metric Tensors of Signature $(2, 2)$

As shown in chapters 2 and 3, the known solutions to the Osserman problem depend on the existence of a solution of the algebraic Osserman problem, that is, the explicit determination of the possible algebraic curvature tensors satisfying the Osserman property.

For semi-Riemannian manifolds with metric tensors of signature $(2, 2)$ the situation is not so simple and, indeed, one may obtain the expressions of all the possible Osserman algebraic curvature tensors and yet may not reach a complete classification of semi-Riemannian Osserman manifolds with metric tensor of signature $(2, 2)$.

The purpose of this section is to study such situation. First of all, in §4.2.1, we will determine the Osserman algebraic curvature tensors on 4-dimensional vector spaces with inner product of neutral signature $(2, 2)$. Section §4.2.2 is devoted to investigate those manifolds which satisfy the Osserman condition at each point, but with eigenvalues possibly changing from point to point. The case when (M, g) is globally Osserman is studied in §4.2.3 and §4.2.4.

4.2.1 Osserman Algebraic Curvature Tensors on Four-Dimensional Vector Spaces of Neutral Signature

In what follows we investigate the existence of certain useful relations for the components of the curvature tensor of a semi-Riemannian manifold (M, g) endowed with a metric tensor of signature $(2, 2)$ which is supposed to be Osserman at a given point $p \in M$. Such relations will play a fundamental role in the explicit determination of the components of the curvature tensor in Theorem 4.2.2.

Fix an orthonormal basis $\{e_1, e_2, e_3, e_4\}$ for T_pM, with $\varepsilon_1 = \varepsilon_2 = -1$ and $\varepsilon_3 = \varepsilon_4 = 1$, where $\varepsilon_i = g(e_i, e_i)$, $i = 1, \ldots, 4$. Since the Osserman condition implies the Einstein property, it is easy to check that the following relations hold.

Lemma 4.2.1. [18] *let (M, g) be a semi-Riemannian manifold with metric tensor of signature $(2, 2)$. If (M, g) is Osserman at $p \in M$, then the components of the curvature tensor at p satisfy*

$$R_{3223} + R_{4224} = R_{3113} + R_{4114}, \quad -R_{2332} + R_{4334} = R_{2112} - R_{4114},$$

$$-R_{2442} + R_{3443} = R_{2112} - R_{3113}, \quad -R_{1331} + R_{4334} = R_{1221} - R_{4224}, \quad (4.9)$$

$$-R_{1441} + R_{3443} = R_{1221} - R_{3223}.$$

$$R_{1221} = R_{4334}, \quad R_{1331} = R_{2442}, \quad R_{1441} = R_{2332}. \quad (4.10)$$

$$R_{3213} + R_{4214} = -R_{2312} + R_{4314} = -R_{2412} + R_{3413},$$
$$-R_{1321} + R_{4324} = -R_{1421} + R_{3423} = 0. \quad (4.11)$$

$$R_{1341} + R_{2342} = 0. \quad (4.12)$$

Note that the above relations are a consequence of the Einstein condition. Next we will derive other relations using the Osserman property. In order to do that, we analyze the traces of the operators R_x and R_x^2 with respect to a vector $x \in T_p M$.

Fixing two indices $i < j$, choose two vectors

$$x = \alpha e_i + \beta e_j, \quad |x|^2 = \alpha^2 \varepsilon_i + \beta^2 \varepsilon_j = \delta, \quad (4.13)$$

and

$$y = \beta \varepsilon_j e_i - \alpha \varepsilon_i e_j, \quad |y|^2 = \varepsilon_i \varepsilon_j \delta, \quad (4.14)$$

such that $\delta \in \{-1, 1\}$ and that $g(x, y) = 0$. From now on we take $\delta = -1$. Next we construct a new orthonormal basis for $T_p M$ say $\{E_k\}$, containing x and y as follows:

$$E_k = \begin{cases} e_k & \text{if} \quad k \neq i, j \\ x & \text{if} \quad k = i \\ y & \text{if} \quad k = j \end{cases} . \quad (4.15)$$

Note that the basis above is orthonormal, and

$$\tilde{\varepsilon}_k = g(E_k, E_k) = \begin{cases} \varepsilon_k & \text{if} \quad k \neq i, j \\ -1 & \text{if} \quad k = i \\ -\varepsilon_i \varepsilon_j & \text{if} \quad k = j \end{cases} .$$

Now consider the Jacobi operator $R_x : (span\{x\})^\perp \to (span\{x\})^\perp$ and write

$$R_x E_k = \sum_{\tilde{s}=1}^{n-1} a_{k\tilde{s}} E_{\tilde{s}},$$

where $a_{\tilde{k}\tilde{l}} = \tilde{\varepsilon}_l g(R_x E_k, E_l)$ and $\tilde{k} = \begin{cases} k & k < i \\ k-1 & k > i \end{cases}$ for $1 \leq k \leq n, k \neq i$. Thus $1 \leq \tilde{k} \leq n-1$ and moreover,

$$a_{k\bar{l}} = \varepsilon_l \left\{ \alpha^2 R_{kiil} + \alpha\beta(R_{kjil} + R_{kijl}) + \beta^2 R_{kjjl} \right\}, \qquad k, l \neq i, j, \qquad (4.16)$$

$$a_{\bar{k}\bar{j}} = \varepsilon_i \varepsilon_j (\beta R_{kjji} - \alpha R_{kiij}), \qquad k \neq i, j, \qquad (4.17)$$

$$a_{\bar{j}\bar{j}} = -\varepsilon_i \varepsilon_j R_{ijji}. \qquad (4.18)$$

Then, it follows that $traceR_x = \sum_{k \neq i} a_{\bar{k}\bar{k}}$ is given by

$$traceR_x = \beta^2 (\rho_{jj} - \varepsilon_i \varepsilon_j \rho_{ii}) + 2\alpha\beta\rho_{ji} - \varepsilon_i \rho_{ii}. \qquad (4.19)$$

Proceeding in an analogous way with the operator R_x^2, one has

$$traceR_x^2 = -2\sigma_2 + (traceR_x)^2, \qquad (4.20)$$

where

$$\sigma_2 = \sum_{k \leq l, \, k, l \neq i} \begin{vmatrix} a_{\bar{k}\bar{k}} & a_{\bar{k}\bar{l}} \\ a_{\bar{l}\bar{k}} & a_{\bar{l}\bar{l}} \end{vmatrix},$$

or

$$\sigma_2 = \sum_{k \neq i} \left(a_{\bar{k}\bar{k}} a_{\bar{j}\bar{j}} - \varepsilon_i \varepsilon_j \varepsilon_k a_{\bar{k}\bar{j}}^2 \right) + \frac{1}{2} \left(\sum_{k \neq i,j} a_{\bar{k}\bar{k}} \right)^2 + \frac{1}{2} \sum_{k,l \neq i,j} \varepsilon_l \varepsilon_k a_{\bar{k}\bar{l}}^2.$$

Using the expressions for $a_{k\bar{l}}$ given by (4.16), (4.17) and (4.18), and also (4.19) for $traceR_x$, a straightforward calculation leads to determining $traceR_x^2$, which has to be independent of β in spite to the Osserman condition. Thus the coefficients of β^4, β^2 and $\alpha\beta^3$ must vanish. Then, after some computations, the following lemma is obtained:

Lemma 4.2.2. [18] *Let (M, g) be a semi-Riemannian manifold with metric tensor of signature $(2, 2)$. If (M, g) is Osserman at $p \in M$, then the components of the curvature tensor at p satisfy*

$$(R_{3124} + R_{3214})^2 + 4R_{1332}^2 = (R_{3113} - R_{3223})^2 + (R_{3114} - R_{3224})^2,$$

$$(R_{2134} + R_{2314})^2 + 4R_{2114}^2 = (R_{3223} + R_{2112})^2 + 4R_{1223}^2,$$

$$(R_{2143} + R_{2413})^2 + 4R_{2113}^2 = (R_{3113} + R_{2112})^2 + 4R_{1224}^2, \qquad (4.21)$$

$$(R_{1234} + R_{1324})^2 + 4R_{1224}^2 = (R_{3113} + R_{2112})^2 + 4R_{2113}^2,$$

$$(R_{1243} + R_{1423})^2 + 4R_{1223}^2 = (R_{2112} + R_{3223})^2 + 4R_{2114}^2.$$

$$(R_{3124} + R_{3214})^2 + 4R_{1332}^2 + R_{2113}^2 + R_{2114}^2 + 2R_{3114}R_{3224}$$
$$= (R_{3113} - R_{3223})^2 + R_{1223}^2 + R_{1224}^2 + 2R_{3114}^2,$$
$$(R_{2134} + R_{2314})^2 + 4R_{2114}^2 + R_{1332}^2 + R_{2113}^2$$
$$= (R_{3223} + R_{2112})^2 + R_{3114}^2 + R_{1334}^2 + 4R_{1223}^2,$$
$$(R_{2143} + R_{2413})^2 + 4R_{2113}^2 + R_{2114}^2 + R_{2441}^2$$
$$= (R_{3113} + R_{2112})^2 + R_{3114}^2 + R_{1223}^2 + 4R_{1224}^2,$$
$$(R_{1234} + R_{1324})^2 + 4R_{1224}^2 + R_{1332}^2 + R_{1223}^2$$
$$= (R_{3113} + R_{2112})^2 + R_{3224}^2 + R_{2114}^2 + 4R_{2113}^2,$$
$$(R_{1243} + R_{1423})^2 + 4R_{1223}^2 + R_{1224}^2 + R_{1442}^2$$
$$= (R_{2112} + R_{3223})^2 + R_{3224}^2 + R_{2113}^2 + 4R_{2114}^2.$$

$$\tag{4.22}$$

$$R_{1332}R_{2114} = -R_{3114}R_{1223}. \tag{4.23}$$

In order to obtain the mentioned description of the Osserman algebraic curvature tensors in indefinite metric tensor signature $(2,2)$, a further observation is needed. As pointed out in the previous section, Jacobi operators in indefinite metric tensor signature $(2,2)$ are not necessarily diagonalizable. Hence, a detailed analysis of the Jordan forms of Jacobi operators is required. This is worked out in the following theorem (see also [114].)

Theorem 4.2.1. [18] *Let (M,g) be a semi-Riemannian manifold with metric tensor of signature $(2,2)$ and let x be a unit vector tangent to M at a point $p \in M$. Then, one of the following holds:*

Type Ia: there exists an orthonormal basis for $(\mathrm{span}\{x\})^\perp$ such that

$$R_x = \begin{pmatrix} \alpha & 0 & 0 \\ 0 & \beta & 0 \\ 0 & 0 & \gamma \end{pmatrix}, \tag{4.24}$$

Type Ib: there exists an orthonormal basis for $(\mathrm{span}\{x\})^\perp$ such that

$$R_x = \begin{pmatrix} \alpha & \beta & 0 \\ -\beta & \alpha & 0 \\ 0 & 0 & \gamma \end{pmatrix}, \tag{4.25}$$

Type II: there exists an orthonormal basis for $(\mathrm{span}\{x\})^\perp$ such that

$$R_x = \begin{pmatrix} \pm(\alpha - \frac{1}{2}) & \pm\frac{1}{2} & 0 \\ \mp\frac{1}{2} & \pm(\alpha + \frac{1}{2}) & 0 \\ 0 & 0 & \beta \end{pmatrix}, \tag{4.26}$$

Type III: there exists an orthonormal basis for $(span\{x\})^\perp$ such that

$$R_x = \begin{pmatrix} \alpha & 0 & \frac{\sqrt{2}}{2} \\ 0 & \alpha & \frac{\sqrt{2}}{2} \\ -\frac{\sqrt{2}}{2} & \frac{\sqrt{2}}{2} & \alpha \end{pmatrix}. \tag{4.27}$$

Note that the different possibilities in the theorem above correspond to the following cases: Jacobi operator R_x is diagonalizable (Type Ia), the characteristic polynomial of Jacobi operators has a complex root (Type Ib), the minimal polynomial of Jacobi operators has a double root (Type II), the minimal polynomial of Jacobi operators has a triple root (Type III.)

Now, using the normal forms of Jacobi operators in previous theorem together with the results in Lemmas 4.2.1 and 4.2.2, the following description of all possible Osserman algebraic curvature tensors is obtained in [18].

Theorem 4.2.2. *Let (M, g) be a semi-Riemannian manifold with metric tensor of signature $(2, 2)$. Then, it is Osserman at $p \in M$ if and only if there exists an orthonormal basis $\{e_1, e_2, e_3, e_4\}$ for $T_p M$ such that the non-vanishing components of the curvature tensor are given by:*

Type Ia:
$$R_{1221} = R_{4334} = \alpha, \quad R_{1331} = R_{4224} = -\beta, \quad R_{1441} = R_{3223} = -\gamma,$$
$$R_{1234} = (-2\alpha + \beta + \gamma)/3, \quad R_{1423} = (\alpha + \beta - 2\gamma)/3, \tag{4.28}$$
$$R_{1342} = (\alpha - 2\beta + \gamma)/3.$$

Type Ib:
$$R_{1221} = R_{4334} = \alpha, \quad R_{1331} = R_{4224} = -\alpha, \quad R_{1441} = R_{3223} = -\gamma,$$
$$R_{2113} = R_{2443} = -\beta, \quad R_{1224} = R_{1334} = \beta, \tag{4.29}$$
$$R_{1234} = (-\alpha + \gamma)/3, \quad R_{1423} = 2(\alpha - \gamma)/3, \quad R_{1342} = (-\alpha + \gamma)/3.$$

Type II:
$$R_{1221} = R_{4334} = \pm\left(\alpha - \tfrac{1}{2}\right), \quad R_{1331} = R_{4224} = \mp\left(\alpha + \tfrac{1}{2}\right),$$
$$R_{1441} = R_{3223} = -\beta,$$
$$R_{2113} = R_{2443} = \mp\tfrac{1}{2}, \quad R_{1224} = R_{1334} = \pm\tfrac{1}{2}, \tag{4.30}$$
$$R_{1234} = \left(\pm\left(-\alpha + \tfrac{3}{2}\right) + \beta\right)/3, \quad R_{1423} = 2(\pm\alpha - \beta)/3,$$
$$R_{1342} = \left(\pm\left(-\alpha - \tfrac{3}{2}\right) + \beta\right)/3.$$

Type III:
$$R_{1221} = R_{4334} = \alpha, \quad R_{1331} = R_{4224} = -\alpha, \quad R_{1441} = R_{3223} = -\alpha,$$
$$R_{2114} = R_{2334} = -\sqrt{2}/2, \quad R_{3114} = -R_{3224} = \sqrt{2}/2, \tag{4.31}$$
$$R_{1223} = R_{1443} = R_{1332} = -R_{1442} = \sqrt{2}/2.$$

Proof. We will study the different possibilities separately. First of all, assume that there is a timelike vector x such that the associated Jacobi operator is of Type Ia. Then, there exists an othonormal basis for T_pM such that R_x has the form (4.24.) As before, we suppose that e_1 and e_2 are timelike ($\varepsilon_1{=}\varepsilon_2{=}{-}1$) and that e_3 and e_4 are spacelike vectors ($\varepsilon_1 = \varepsilon_2 = 1$.) (Also we continue with $\delta = -1$ in (4.13)). Next we compute the components of the curvature tensor R.

Denote by α, β and γ the eigenvalues of R_{e_1} corresponding to the eigenvectors e_2, e_3 and e_4, respectively. First, one easily checks that

$$R_{3443} = R_{2112} = \alpha, \quad R_{2332} = R_{4114} = -\gamma, \quad R_{4224} = R_{3113} = -\beta. \quad (4.32)$$

To compute R_{k11l} for $k \neq l$, we use that e_2, e_3 and e_4 are eigenvectors of R_{e_1} with corresponding eigenvalues α, β and γ, respectively, and hence

$$R_{k11l} = 0, \quad \text{for } k \neq l \text{ and } k, l = 2, 3, 4. \quad (4.33)$$

In the next step, one obtains

$$R_{kttl} = 0, \quad \text{for } t \neq k \neq l \text{ and } t, k, l = 1, 2, 3, 4, \quad (4.34)$$

by combining (4.22) and (4.23) with the components in (4.32) and (4.33), and also considering the relations in (4.9) and (4.11).

Finally, it remains to determine R_{1234} and R_{1423}, and then $R_{1342} = -(R_{1234} + R_{1423})$. Put $R_{1234} = x$ and $R_{1423} = y$. The first relation in (4.23) together with (4.32) yields

$$(x + 2y)^2 = (\beta - \gamma)^2. \quad (4.35)$$

Analogously, the second and third relations in (4.22) imply

$$(y - x)^2 = (\gamma - \alpha)^2 \quad (4.36)$$

and

$$(2x + y)^2 = (\alpha - \beta)^2, \quad (4.37)$$

respectively. If α, β and γ are different, the solution of the system determined by (4.35)-(4.37) is given by

$$\begin{cases} x = \dfrac{-2\alpha + \beta + \gamma}{3}, \\ y = \dfrac{\alpha + \beta - 2\gamma}{3}. \end{cases} \quad (4.38)$$

(There is also negative of the solution above, but using the transformation $e_1 \mapsto -e_1$ and $e_i \mapsto e_i$ for $i = 2, 3, 4$, we can consider only the solution in (4.38)). If α, β and γ are not different, it can be directly checked that the solution also corresponds to (4.38). Thus, all the components of the curvature tensor are determined.

Next, we assume that there is a timelike unit vector x such that the associated Jacobi operator is of type Ib, that is, there is an orthonormal basis for T_pM such that R_x is given by (4.25). Proceeding as before, one has

$$R_{1221} = R_{4334} = \alpha, \quad R_{1331} = R_{4224} = -\alpha, \quad R_{1441} = R_{3223} = -\gamma,$$

$$R_{2113} = R_{2443} = -\beta, \quad R_{1224} = R_{1334} = \beta,$$

$$R_{1234} = x = (-\alpha + \gamma)/3, \quad R_{1423} = y = 2(\alpha - \gamma)/3, \tag{4.39}$$

$$R_{1342} = -x - y,$$

with

$$x - y = \gamma - \alpha, \quad 2x + y = 0, \quad x + 2y = \alpha - \gamma,$$

and the others being zero.

In order to show that (4.30) holds, let us assume the existence of a timelike unit vector whose associated Jacobi operator has the form

$$R_x = \begin{pmatrix} \mp(\alpha - \frac{1}{2}) & \pm\frac{1}{2} & 0 \\ \mp\frac{1}{2} & \mp(\alpha + \frac{1}{2}) & 0 \\ 0 & 0 & -\beta \end{pmatrix}. \tag{4.40}$$

Then, proceeding as in the previous cases, the nonzero components of the curvature tensor are given by

$$R_{1221} = R_{4334} = \pm(\alpha - \frac{1}{2}), \quad R_{1331} = R_{4224} = \mp(\alpha + \frac{1}{2}),$$

$$R_{1441} = R_{3223} = -\beta,$$

$$R_{2113} = R_{2443} = \mp\frac{1}{2}, \quad R_{1224} = R_{1334} = \varepsilon \pm \frac{1}{2}, \tag{4.41}$$

$$R_{1234} = x = (\pm(-\alpha + \frac{3}{2}) + \beta)/3, \quad R_{1423} = y = 2(\pm\alpha - \beta)/3,$$

$$R_{1342} = -x - y,$$

with

$$x - y = \beta \mp (\alpha - \frac{1}{2}), \quad 2x + y = \pm 1, \quad x + 2y = \pm(\alpha + \frac{1}{2}) - \beta, \tag{4.42}$$

where $\varepsilon = \pm 1$ depending on the choice of the orientation for the orthonormal basis of T_pM (just replacing e_1 by $-e_1$.)

Finally, if there exists a timelike unit vector x such that R_x is Type III, proceeding as in the previous case, (4.31) is obtained. □

Remark 4.2.1. The previous theorem classifies all possible Osserman algebraic curvature tensors on a 4-dimensional vector space with an inner product of signature $(2, 2)$. Now, proceeding as in Remark 1.2.5, one may construct metric tensors of signature $(2, 2)$ which are Osserman at a point and moreover, all the possibilities for the Jacobi operators in Theorem 4.2.1 are also realized at that point.

This should be contrasted with the fact that a complete classification of the semi-Riemannian Osserman manifolds with metric tensors of signature $(2, 2)$ is not yet available. Moreover, note that Type Ib semi-Riemannian Osserman manifolds do not exist (cf. Theorem 4.2.10.)

It is clear from the results in the previous section that the characteristic and minimal polynomials of Jacobi operators play different roles in semi-Riemannian geometry. Indeed, the minimal polynomials may have nonconstant roots, counting multiplicities, even if the manifold is Osserman (cf. Remark 4.1.2.) Therefore, it seems natural to investigate those Osserman manifolds where the roots of the minimal polynomials of Jacobi operators are constant, counting multiplicities. This is the approach mainly developed by Blažić, Bokan, and Rakić [18], who investigated the geometry of such manifolds. We recall the following:

Definition 4.2.1. [18] *Let (M, g) be a semi-Riemannian manifold.*

a) *(M, g) is called Jordan-Osserman at a point $p \in M$ if the Jordan form of R_z is independent of $z \in S_p(M)$.*

b) *(M, g) is called pointwise Jordan-Osserman if it is Jordan-Osserman at each point $p \in M$.*

c) *(M, g) is called globally Jordan-Osserman if the Jordan form of R_z is independent of $z \in S(M)$.*

Remark 4.2.2. The possible relations between the pointwise and global Jordan-Osserman conditions are not well-understood as in the corresponding ones between the pointwise and global Osserman conditions. In fact, a result similar to the one in Theorem 1.3.1 should not be expected. Indeed, the examples given in Remark 4.1.2 are all pointwise Jordan-Osserman, since the roots of minimal polynomials are independent of the direction at each point. However multiplicity of such roots may change from point to point, in showing that they are not globally Jordan-Osserman although Jacobi operators have only one eigenvalue.

Note that a semi-Riemannian pointwise Osserman manifold with metric tensor of signature $(2, 2)$ is pointwise Jordan-Osserman. This is a consequence of Theorem 4.2.2. In fact, a straightforward calculation shows that if (M, g) is Osserman at $p \in M$ and R_x is diagonalizable (resp., Types Ib, II, III) for some unit vector $x \in T_pM$, then R_z is diagonalizable (resp., Types Ib, II, III) for all unit vectors z in T_pM. However, the situation is different for higher dimensions (cf. Remark 5.1.1.)

Note that any Riemannian Osserman manifold is automatically Jordan-Osserman since the Jacobi operators are always diagonalizable for positive definite metric tensors.

We recall here some basic facts related to the local homogeneity of a semi-Riemannian manifold. Let $\nabla^i R(p)$ denote the ith covariant differential of the curvature tensor at a point $p \in M$. Then (M, g) is called *curvature homogeneous up to order* $k \in \mathbb{N}$ [131] if for any pair of points $p, q \in M$, there is a linear isometry $F : T_p M \to T_q M$ such that

$$F^*(\nabla^i R(q)) = \nabla^i R(p), \qquad \text{for all } i = 0, \ldots, k. \qquad (4.43)$$

Note that any locally homogeneous semi-Riemannian manifold is curvature homogeneous up to any order k. The converse is also true. In fact, it suffices to check the curvature homogeneity up to certain order k_M (see [131], [136].)

Note that any semi-Riemannian pointwise (Jordan-) Osserman manifold which is curvature homogeneous up to order zero is globally (Jordan-) Osserman. Conversely, we have,

Theorem 4.2.3. [23] *Let* (M, g) *be a Jordan-Osserman manifold with metric tensor of signature* $(2, 2)$. *Then it is curvature homogenous up to order zero.*

Proof. Since (M, g) is assumed to be Jordan-Osserman, the Jordan form of the Jacobi operators does not change from one point to another and the result follows from Theorem 4.2.2.   □

It is still an open question whether or not Riemannian Osserman manifolds are locally homogeneous, although a positive answer could be expected in spite of the results in [122].

Remark 4.2.3. Note that semi-Riemannian Jordan-Osserman manifolds are not necessarily locally homogeneous. In fact, it is shown in [23] that the metric tensors $g_{(f_1, f_2)}$ on $\mathbb{R}^4$ defined by

$$\begin{aligned} g_{(f_1, f_2)} &= f_1(x_1, x_2)^2 dx_1 \otimes dx_1 + f_2(x_1, x_2)^2 dx_2 \otimes dx_2 \\ &+ dx_1 \otimes dx_3 + dx_3 \otimes dx_1 + dx_2 \otimes dx_4 + dx_4 \otimes dx_2 \end{aligned} \qquad (4.44)$$

are of Type II Jordan-Osserman if

$$\frac{\partial^2}{\partial x_2^2} f_1(x_1, x_2)^2 + \frac{\partial^2}{\partial x_1^2} f_2(x_1, x_2)^2 > 0. \qquad (4.45)$$

Moreover, such metric tensors are not locally homogeneous if

$$\frac{64}{\left(\frac{\partial^2 f_1^2}{\partial x_2^2} + \frac{\partial^2 f_2^2}{\partial x_1^2}\right)^5} \left(\frac{\partial^3 f_1^2}{\partial x_2^3} + \frac{\partial^3 f_2^2}{\partial x_1^2 \partial x_2}\right)^2 \left(\frac{\partial^3 f_1^2}{\partial x_2^2 \partial x_1} + \frac{\partial^3 f_2^2}{\partial x_1^3}\right)^2 \neq const. \qquad (4.46)$$

Now, one can easily check that $(\mathbb{R}^4 \backslash \{0\}, g_{(f_1, f_2)})$, where $f_1(x_1, x_2) = f_2(x_1, x_2) = x_1 x_2$, satisfies (4.45) and thus $(\mathbb{R}^4 \backslash \{0\}, g_{(f_1, f_2)})$ is a Type II Jordan-Osserman manifold. However (4.46) does not hold, and hence $(\mathbb{R}^4 \backslash \{0\}, g_{(f_1, f_2)})$ is not locally homogeneous.

Although a Jordan-Osserman manifold with metric tensor of signature $(2, 2)$ is not necessarily locally homogeneous (cf. Remark 4.2.3), and therefore not locally symmetric, it still satisfies the following relation.

Theorem 4.2.4. [23] *Let (M, g) be a semi-Riemannian Jordan-Osserman manifold with metric tensor of signature $(2, 2)$. Then ∇R is isotropic, that is, $\|\nabla R\|^2 = 0$.*

4.2.2 Four-Dimensional Semi-Riemannian Pointwise Osserman Manifolds

In this subsection, we investigate further the existence of semi-Riemannian pointwise Osserman manifolds with metric tensors of signature $(2, 2)$. In order to construct such examples, we show in Theorem 4.2.5 the equivalence between the Osserman condition and Einstein self-duality (or Einstein anti-self-duality) at a given point $p \in M$. As an application, many new examples of semi-Riemannian pointwise Osserman manifolds are derived from those in [87], [88], [120].

We start with some basic material and definitions.

Let (V, g) be an n-dimensional inner product space with inner product of signature (ν, η) and let v_g denote its volume element. Then $\bigwedge^2 V^*$ can be endowed with the inner product, also denoted by g, given by the formula

$$g(e^i \wedge e^j, e^k \wedge e^l) = \det \begin{pmatrix} g(e_i, e_k) & g(e_i, e_l) \\ g(e_j, e_k) & g(e_j, e_l) \end{pmatrix}.$$

Define the Hodge star operator $*$ to be the isomorphism $* : \bigwedge^k V^* \to \bigwedge^{n-k} V^*$ by $\alpha \wedge *\beta = g(\alpha, \beta) v_g$ for all k-forms $\alpha, \beta \in \bigwedge^k V^*$. Then this operator satisfies $*^2 = (-1)^{k(n-k)+\eta} id_{\bigwedge^k V^*}$. Hence, for dimension $n = 4$, $*$ induces an automorphism of $\bigwedge^2 V^*$, which is an involution if the inner product is positive definite or indefinite of signature $(2, 2)$. Moreover, $*$ defines a complex structure on $\bigwedge^2 V^*$ if the inner product is of signature $(1, 3)$.

We restrict our attention to the cases when $*^2 = id_{\bigwedge^2 V^*}$. Then $\bigwedge^2 V^*$ decomposes as $\bigwedge^2 V^* = \bigwedge_+ \oplus \bigwedge_-$, where $\bigwedge_+$ and $\bigwedge_-$ are the eigenspaces corresponding to the eigenvalues $+1$ and -1 of $*$, respectively, where $\bigwedge_\pm = \{\alpha \in \bigwedge^2 V^*; *\alpha = \pm\alpha\}$. The 2-forms in $\bigwedge_+$ and $\bigwedge_-$ are usually called *self-dual* and *anti-self dual*, respectively. Furthermore, note that the inner product on $\bigwedge^2 V^*$ is positive definite and indefinite of signature $(3, 3)$ if (V, g)

is positive definite or indefinite of signature $(2, 2)$, respectively. Moreover, in the latter case, $\bigwedge_{\pm}$ are Lorentzian subspaces of $(\bigwedge^2 V^*, g)$.

Now, consider an orthonormal basis $\{e_1, e_2, e_3, e_4\}$ for (V, g) such that e_1, e_2 are timelike unit vectors and let $\{e^1, e^2, e^3, e^4\}$ denotes the corresponding dual basis. Then

$$\bigwedge_+ = \text{span } \{ (e^1 \wedge e^2 + e^3 \wedge e^4)/\sqrt{2}, (e^1 \wedge e^3 - e^4 \wedge e^2)/\sqrt{2},$$
$$(e^1 \wedge e^4 - e^2 \wedge e^3)/\sqrt{2} \},$$

and

$$\bigwedge_- = \text{span } \{ (e^1 \wedge e^2 - e^3 \wedge e^4)/\sqrt{2}(e^1 \wedge e^3 + e^4 \wedge e^2)/\sqrt{2},$$
$$(e^1 \wedge e^4 + e^2 \wedge e^3)/\sqrt{2} \}.$$

Let F be an algebraic curvature map on (V, g), and denote by Ric^F and Sc^F its Ricci tensor and scalar curvature, respectively. Moreover, let Z^F denote the trace-free Ricci tensor, that is,

$$Z^F(x, y) = Ric^F(x, y) - \frac{Sc^F}{4} g(x, y) \qquad \text{for all } x, y \in V.$$

Then, the Weyl tensor W^F of F becomes (see 3.1)

$$W^F(x_1, x_2, x_3, x_4) = F(x_1, x_2, x_3, x_4) - \frac{Sc^F}{12} R^0(x_1, x_2, x_3, x_4)$$
$$- \frac{1}{2} \{ g(x_1, x_4) Z^F(x_2, x_3) - g(x_2, x_4) Z^F(x_1, x_3)$$
$$+ g(x_2, x_3) Z^F(x_1, x_4) - g(x_1, x_3) Z^F(x_2, x_4) \},$$

where $x_1, x_2, x_3, x_4 \in V$.

Considering F, Z^F and W^F as linear maps on $\bigwedge^2 V^*$, it can be shown that $*W^F = W^F*$ and $*Z^F = -Z^F*$. Hence, the algebraic curvature map F can be decomposed as:

$$R = (Sc^F/12) id_{\bigwedge^2 V^*} \oplus Z^F \oplus (W_+^F \oplus W_-^F), \qquad (4.47)$$

where $W_{\pm}^F$ are the restrictions of W^F to $\bigwedge_{\pm}$.

Now, for a positive definite or an indefinite inner product of signature $(2, 2)$, we have,

Definition 4.2.2. *Let (V, g) be an inner product space and F be an algebraic curvature map on (V, g). Then (V, g, F) is called self-dual (resp., anti-self-dual) if $W_-^F \equiv 0$ (resp., $W_+^F \equiv 0$.)*

A 4-dimensional semi-Riemannian manifold (M, g) is called self-dual (resp., anti-self-dual) if it is oriented and its curvature tensor is self-dual (resp., anti-self-dual) at each point $p \in M$ (see, for example, [87] and the references therein.)

In what follows, we investigate the relationship between the self (resp., anti-self)-duality and the Osserman condition. It is important to note here that, because of the indefiniteness of the inner product on $\bigwedge_{\pm}$, the self-dual and anti-self-dual parts of the curvature tensor are not necessarily diagonalizable. Furthermore, if (M, g) is assumed to be self-dual, it is called Type Ia, Ib, II or III if W_+ is diagonalizable, has a complex eigenvalue, has a double root of its minimal polynomial or has a triple root of its minimal polinomial, respectively. Now we have the following.

Theorem 4.2.5. [3] *Let (M, g) be a semi-Riemannian manifold with metric tensor of signature $(2, 2)$. Then, the following conditions are equivalent:*

1. *(M, g) is pointwise Osserman.*
2. *There is a choice of orientation for M such that it is Einstein self-dual (or Einstein anti-self-dual.)*

Moreover, the types of the self-dual curvature W_+ (resp., anti-self-dual curvature W_-) are in one to one correspondence with the different types of the Jacobi operators in Theorem 4.2.1 .

Proof. If (M, g) is pointwise Osserman, then it is Einstein and moreover, the Jacobi operators is of one of the forms in (4.24), (4.25) (4.26) or (4.27). At a given point $p \in M$, take an orthonormal basis $\{e_1, e_2, e_3, e_4\}$ for $T_p M$ where the components of the curvature tensor are expressed as in Theorem 4.2.2. Then, after a straightforward computation, it can be shown that the curvature operator on $\bigwedge^2 T_p^* M$, satisfies:

$$R = \frac{Sc}{12} id_{\bigwedge^2(T_p^* M)} + \begin{pmatrix} W_+ & 0 \\ 0 & 0 \end{pmatrix},$$

where W_+ is given by one of the following:

$$W_+ = \begin{pmatrix} \frac{1}{2}\alpha & 0 & 0 \\ 0 & \frac{1}{2}\beta & 0 \\ 0 & 0 & -\frac{1}{2}(\alpha_1 + \alpha_2) \end{pmatrix}$$

if (M, g) is Type Ia Osserman at $p \in M$, or

$$W_+ = \begin{pmatrix} \frac{1}{2}\alpha & -\frac{1}{2}\beta & 0 \\ \frac{1}{2}\beta & \frac{1}{2}\alpha & 0 \\ 0 & 0 & -\alpha \end{pmatrix}$$

if (M, g) is Type Ib Osserman at $p \in M$, or

$$W_+ = \begin{pmatrix} \frac{1}{2}\alpha - \frac{1}{4} & \frac{1}{4} & 0 \\ -\frac{1}{4} & \frac{1}{2}\alpha + \frac{1}{4} & 0 \\ 0 & 0 & -\alpha \end{pmatrix}$$

if (M, g) is Type II Osserman at $p \in M$, or

$$W_+ = \begin{pmatrix} 0 & 0 & -\frac{1}{2\sqrt{2}} \\ 0 & 0 & -\frac{1}{2\sqrt{2}} \\ \frac{1}{2\sqrt{2}} & -\frac{1}{2\sqrt{2}} & 0 \end{pmatrix}$$

if (M, g) is Type III Osserman at $p \in M$.

This shows that the semi-Riemannian pointwise Osserman manifolds with metric tensors of signature $(2, 2)$ and with a suitable orientation are Einstein self-dual or Einstein anti-self-dual and moreover, the different types of the Jacobi operators are in one to one correspondence with different types of $W_\pm$.

Next we show its converse. Let $p \in M$ and assume that (M, g) is Einstein self-dual at T_pM. Then, it follows from (4.47) that

$$R = (Sc/12)id_{\wedge^2(T_p^*M)} + \begin{pmatrix} W_+ & 0 \\ 0 & 0 \end{pmatrix}. \tag{4.48}$$

Now, we consider separately the different possibilities for W_+.

a)   If W_+ is diagonalizable with eigenvalues α_1, α_2 and $\alpha_3 = Sc - \alpha_1 - \alpha_2$, then there exists an orthonormal basis $\{e_1, \ldots, e_4\}$ for T_pM such that $W_+ = diag[\alpha_1, \alpha_2, \alpha_3]$ in the induced basis $\{(e^1 \wedge e^2 + e^3 \wedge e^4)/\sqrt{2}, (e^1 \wedge e^3 - e^4 \wedge e^2)/\sqrt{2}, (e^1 \wedge e^4 - e^2 \wedge e^3)/\sqrt{2}\}$ for W_+. Then, the components of the curvature tensor at T_pM with respect to $\{e_1, \ldots, e_4\}$ are given by

$R_{1221} = R_{4334} = -(1/2)\alpha_1 - (Sc/12)$,

$R_{1331} = R_{4224} = (1/2)\alpha_2 + (Sc/12)$,

$R_{1441} = R_{3223} = -(1/2)(\alpha_1 + \alpha_2) + (Sc/12)$,

$R_{1234} = (1/2)\alpha_1$, $R_{1342} = (1/2)\alpha_2$, $R_{1423} = -(1/2)(\alpha_1 + \alpha_2)$.

Hence, it follows from (4.28) in Theorem 4.2.2 that (M, g) is Osserman Type Ia with eigenvalues $\alpha = -\frac{1}{2}\alpha_1 - \frac{Sc}{12}$, $\beta = -\frac{1}{2}\alpha_2 - \frac{Sc}{12}$ and $\gamma = \frac{1}{2}(\alpha_1 + \alpha_2) + \frac{Sc}{12}$.

b)   If W_+ has a complex eigenvalue $(\frac{\alpha}{2} + i\frac{\gamma}{2})$ and a real eigenvalue $-\alpha$, then it follows from (4.48) that the nonvanishing components of the curvature tensor at $p \in M$ are given by

$R_{1221} = R_{4334} = -R_{1331} = -R_{4224} = -(1/2)\alpha - (Sc/12)$,

$R_{1441} = R_{3223} = -\alpha + (Sc/12)$,

$R_{1234} = R_{1342} = (1/2)\alpha$, $R_{1423} = -\alpha$,

$R_{2113} = -R_{1224} = -R_{1334} = R_{3442} = (1/2)\gamma$, and thus, the characteristic polynomial of the Jacobi operator R_z is given by

$$p_\lambda(R_z) = \left[\lambda - (\alpha - \frac{Sc}{12})\varepsilon_z\right][\lambda + \xi\varepsilon_z]\left[\lambda + \bar{\xi}\varepsilon_z\right],$$

where $\xi = (\frac{\alpha}{2} + \frac{Sc}{12}) - i\frac{\gamma}{2} \in \mathbb{C}$ and $z \in T_pM$ is a unit vector. Hence (M, g) is Osserman of Type Ib at $p \in M$.

c)   Next, assume the minimal polynomial of W_+ has a double root $\frac{\alpha}{2}$ and a simple root $-\alpha$. Then, the nonvanishing components of the curvature tensor at $p \in M$ are given by

$R_{1221} = R_{4334} = -(1/2)\alpha + (1/4) - (Sc/12),$

$R_{1331} = R_{4224} = (1/2)\alpha + (1/4) + (Sc/12),$

$R_{1441} = R_{3223} = -\alpha + (Sc/12),$

$R_{1234} = (1/2)\alpha - (1/4), \quad R_{1342} = (1/2)\alpha + (1/4), \quad R_{1423} = -\alpha,$

$R_{2113} = -R_{1224} = -R_{1334} = R_{3442} = -1/4.$
   Then, for a unit vector $z \in T_pM$, we have

$$p_\lambda(R_z) = m_\lambda(R_z) = \left[\lambda + (\frac{\alpha}{2} + \frac{Sc}{12})\varepsilon_z\right]^2\left[\lambda - (\alpha - \frac{Sc}{12})\varepsilon_z\right],$$

in showing that (M, g) is Osserman of Type II at $p \in M$, where $p_\lambda(R_z)$ and $m_\lambda(R_z)$ are the characteristic and minimal polynomials of R_z, respectively.

d)   Finally, if the minimal polynomial of W_+ has a triple root, then the nonvanishing components of the curvature tensor at $p \in M$ are given by
$R_{1221} = R_{4334} = -R_{1331} = -R_{4224} = -R_{1441} = -R_{3223} = -Sc/12,$

$R_{2114} = R_{1223} = R_{1443} = R_{2334} = 1/(2\sqrt{2}),$

$R_{3114} = -R_{1332} = R_{1442} = -R_{4223} = -1/(2\sqrt{2}).$
   Then, the characteristic and minimal polynomials of the Jacobi operator R_z are given by

$$p_\lambda(R_z) = m_\lambda(R_z) = \left[\lambda + \frac{Sc}{12}\varepsilon_z\right]^3,$$

for any unit vector $z \in T_pM$. Hence the manifold is Osserman of Type III at p.   □

Remark 4.2.4. The equivalence between Einstein self (resp., anti-self)-duality and pointwise Osserman condition in Riemannian geometry (cf. Theorem 2.1.8) is obtained in a similar way as in the previous theorem for Type Ia manifolds.

Remark 4.2.5. As a further application of Theorem 4.2.5, an example of a *compact* semi-Riemannian Osserman manifold is obtained as follows [120].

On the torus $M = \mathbb{C}/\Gamma_1 \times \mathbb{C}/\Gamma_2$, consider $\gamma = f(z)dz \wedge d\bar{z} + dz \wedge d\bar{w} + dw \wedge d\bar{z}$ where z and w are holomorphic coordinates on each complex plane and f is a smooth real positive valued function on $\mathbb{C}\Gamma_1$. It is clear that γ defines an indefinite Kähler metric tensor g on M.

Now, considering the curvature tensor as a section of $End(\bigwedge^2(T^*M))$ and taking into account the splitting $\bigwedge^2(T^*M) = \bigwedge_+(T^*M) \oplus \bigwedge_-(T^*M)$, we have

$$R = \begin{pmatrix} A & B \\ B^* & D \end{pmatrix} \tag{4.49}$$

and moreover, $B = D = 0$ and, with respect to a basis for $\bigwedge_+$,

$$A = (2f^{-2}\Delta f) \begin{pmatrix} 1 & 0 & -1 \\ 0 & 0 & 0 \\ 1 & 0 & -1 \end{pmatrix}. \tag{4.50}$$

This shows that the metric tensor is Einstein and self-dual. Therefore, it is pointwise Osserman. Moreover, it follows from (4.49) and (4.50) that the eigenvalues of the Jacobi operators are zero, and thus it is Osserman.

4.2.3 Symmetric Semi-Riemannian Osserman Manifolds with Metric Tensors of Signature $(2, 2)$

When restricting our attention to the class of locally symmetric spaces, it follows that locally symmetric Riemannian Osserman manifolds are either flat or locally isometric to a rank-one symmetric space (cf. Theorem 2.2.5.) The same conclusion holds too for any locally symmetric Lorentzian Osserman manifolds (cf. Theorem 3.1.2.) However, the analogous result is no longer true in the geometry of metric tensors of signature $(2, 2)$, as pointed out in Remark 4.1.4. Moreover, note that an example of a rank-two symmetric semi-Riemannian Osserman manifold with metric tensor of signature $(2, 2)$ was constructed by Rakić in [123]. The purpose of this section is to classify the locally symmetric semi-Riemannian Osserman manifolds with metric tensors of signature $(2, 2)$ as a first step towards the more general study of semi-Riemannian Jordan-Osserman manifolds.

Since any semi-Riemannian Osserman manifold is Einstein (cf. Proposition 1.2.1), we may restrict our attention to Einstein manifolds with metric tensors of signature $(2, 2)$. The class of locally symmetric semi-Riemannian Einstein manifolds with metric tensors of signature $(2, 2)$ contains the obvious examples provided by the real space forms, complex space forms, and products of two surface metric tensors with the same constant Gaussian curvatures. Although those are the only possibilities for positive definite metric tensors, some other examples exist for indefinite metric tensors. Such manifolds are discussed in detail by Derdzinski [47]. Next, we recall the construction of two of such examples when the metric tensor is of signature $(2, 2)$.

Example 4.2.1. In full analogy with real-analytic semi-Riemannian metric tensors on real manifolds, one can speak of *complex-analytic metric tensors on complex manifolds.* Let M be a complex manifold of some complex dimension n. By a *complex-analytic metric tensor* g on M we mean the assignment to each point $p \in M$ a nondegenerate complex-bilinear symmetric form $g(p)$: $T_pM \times T_pM \to \mathbb{C}$ whose dependence on p is complex-analytic. Since our discussion is local, we may as well fix a complex-analytic local coordinate system z_j in M, $j = 1, \ldots, n$, thus identifying the coordinate domain with a region Ω in $\mathbb{C}^n$. A complex-analytic metric tensor g on Ω is now described by its component functions g_{ij} with $g_{ij} = g(e_i, e_j)$, where e_k, $k = 1, \ldots, n$, are the vectors in the standard basis for $\mathbb{C}^n$, treated as parallel vector fields. The requirements of complex-analyticity, nondegeneracy, and symmetry in the above definition now mean, respectively, that g_{ij} are all complex-analytic, while $\det(g_{ij}) \neq 0$ and $g_{ij} = g_{ji}$ at every point of Ω.

Any complex-analytic metric tensor g has a well-defined Levi-Civita connection ∇, curvature tensor R, Ricci tensor Ric, and scalar curvature function Sc, all defined by the same local-coordinate formula as in the real case. The only difference lies in the required regularity: Since the operators $\frac{\partial}{\partial z_j}$ now are the complex (Cauchy-Riemann) partial derivatives, all functions they are applied to, or resulted from their application, must be complex-analytic. As in the coordinate-free meaning of these objects, it is completely analogous to that for real metric tensors. In particular, we may also speak of complex-analytic metric tensors which are locally symmetric or Einstein.

Complex-analytic metric tensors give rise to very easy constructions of (real) semi-Riemannian Einstein metric tensors. The real part $g = Re\,h$ of any complex-analytic metric tensor h in the complex dimension n is a (real) semi-Riemannian metric in the real dimension $2n$; in fact, g is nondegenerate and moreover, g must have neutral signature (n, n) since $g(iv, iw) = -g(v, w)$ for tangent vectors v, w. (Multiplication by i establishes, at any point, an algebraic equivalence between g and $-g$.)

If we start with a (real) surface metric tensor h having nonzero constant Gaussian curvature and form its "local complexification", or *complex-analytic extension*, which is a complex-analytic metric tensor $h^{\mathbb{C}}$ on a complex surface, then its real part $g = Re\,h^{\mathbb{C}}$ is a real locally symmetric Einstein semi-Riemannian metric tensor. In addition, its Weyl tensor W satisfies $W_+ \neq 0$, $W_- \neq 0$ at every point. Due to the uniqueness of analytic continuation, all relations valid for the original metric tensor h that appear in local coordinates as polynomial equalities involving h_{jk} and their partial derivatives up to any given order are also valid for $h^{\mathbb{C}}$. Thus, for instance, $h^{\mathbb{C}}$ is locally symmetric, or Einstein, if so is h.

Example 4.2.2. Suppose that we are given any symmetric 2×2 real matrix $\mathfrak{G}$ with $\det \mathfrak{G} \neq 0$ and any C^∞ function f of two variables x_1, x_2, defined on an open subset of $\mathbb{R}^2$. Let $\{x_j\}$ be a coordinate system on a 4-dimensional manifold, and let $\frac{\partial}{\partial x_j}$, $j = 1, \ldots, 4$, be the corresponding coordinate vector fields. We now define an indefinite metric tensor g on the coordinate domain by requiring its component functions g_{jk} to form the block matrix

$$[g_{jk}] = \begin{bmatrix} \mathfrak{G} & 0 \\ 0 & \mathfrak{M} \end{bmatrix}$$

consists of 2×2 matrices, where

$$\mathfrak{M} = \begin{bmatrix} 0 & 1 \\ 1 & -f \end{bmatrix}$$

with $f = f(x_1, x_2)$ treated as a function of (x_1, x_2, x_3, x_4).

This construction leads to many examples of Ricci-flat metric tensors. Moreover, the locally symmetric Einstein metric tensors of signature $(2,2)$ constructed by this procedure can be described in suitable coordinates (x_1, x_2, x_3, x_4) by

$$g_f = dx_1 \otimes dx_2 + dx_2 \otimes dx_1 + dx_3 \otimes dx_4 + dx_4 \otimes dx_3 - f(x_1, x_2)dx_4 \otimes dx_4.$$

where f is one of the following

$$f = \pm(x_1)^2, \quad f = \pm[(x_1)^2 + (x_2)^2], \quad f = (x_1)^2 - (x_2)^2.$$

Now, one has the following useful local description of locally symmetric Einstein manifolds with metric tensors of signature $(2,2)$.

Theorem 4.2.6. [47] *Let (M, g) be a locally symmetric Einstein manifold with a metric tensor g of signature $(2,2)$. Then, only one of the following cases occurs.*

(i) *(M, g) is a real space form,*
(ii) *(M, g) is a nonflat complex space form,*
(iii) *(M, g) is a nonflat paracomplex space form,*
(iv) *The metric g is, locally, a product of two semi-Riemannian surface metric tensors with signature $(2, 0)$ and $(0, 2)$, having equal nonzero constant Gaussian curvatures,*
(v) *The metric g is, locally, a product of two semi-Riemannian surface metric tensors with equal nonzero constant Gaussian curvatures, both are of signature $(1, 1)$*
(vi) *(M, g) is, locally, the result of complexifying a positive-definite surface metric tensor with a nonzero constant Gaussian curvature,*
(vii) *The metric g is locally the same with one of the five metric tensors described in Example 4.2.2.*

On the basis of the previous theorem, one has the following:

Theorem 4.2.7. *Let (M, g) be a nonflat 4-dimensional locally symmetric semi-Riemannian Osserman manifold. Then one of the following holds.*

(i) (M, g) is locally isometric to a rank-one symmetric space, or
(ii) (M, g) is locally isometric to the rank-two symmetric space G/H, where $\{\omega^0, \omega^1, \omega^2, \omega^3, \omega^4\}$ with

$$d\omega^0 = d\omega^1 = 0,$$

$$d\omega^2 = -\omega^0 \wedge \omega^1, \quad d\omega^3 = \omega^1 \wedge \omega^2, \quad d\omega^4 = -\omega^0 \wedge \omega^2$$

form a basis for $\mathfrak{g}^*$, where $\mathfrak{g} = Lie(G)$ and $\mathfrak{h} = Lie(H)$ is given by $\mathfrak{h}^* = span\{\omega^2\}$, or
(iii) (M, g) is locally isometric to the rank-two symmetric space $\tilde{G}/\tilde{H}$, where $\{\tilde{\omega}^0, \tilde{\omega}^1, \tilde{\omega}^2, \tilde{\omega}^3, \tilde{\omega}^4\}$ with

$$d\tilde{\omega}^0 = d\tilde{\omega}^1 = 0,$$

$$d\tilde{\omega}^2 = -\tilde{\omega}^0 \wedge \tilde{\omega}^1, \quad d\tilde{\omega}^3 = -\tilde{\omega}^1 \wedge \tilde{\omega}^2, \quad d\tilde{\omega}^4 = \tilde{\omega}^0 \wedge \tilde{\omega}^2$$

form a basis for $\tilde{\mathfrak{g}}^*$, where $\tilde{\mathfrak{g}} = Lie(\tilde{G})$ and $\tilde{\mathfrak{h}} = Lie(\tilde{H})$ is given by $\tilde{\mathfrak{h}}^* = span\{\tilde{\omega}^2\}$.

Proof. Note that any Osserman manifold with metric tensor of signature $(2, 2)$ is Einstein self-dual or anti-self dual. Therefore, it immediately follows from the different cases in previous theorem that (M, g) is locally isometric to one of the following:

(a) a real space form,
(b) a nonflat complex space form,
(c) a nonflat paracomplex space form,
(d) (M, g) is locally isometric to $\mathbb{R}^4$ with a metric tensor

$$g_{\pm} = dx_1 \otimes dx_2 + dx_2 \otimes dx_1 + dx_3 \otimes dx_4 + dx_4 \otimes dx_3 \mp (x_1)^2 dx_4 \otimes dx_4.$$

A straightforward computation now shows that the nonvanishing components of the curvature tensors of those manifolds corresponding to case (d) above are given by

$$R(e_1, e_4)e_1 = -e_3$$
$$R(e_1, e_4)e_4 = e_2 \tag{4.51}$$

if $f = (x_1)^2$, and by

$$\tilde{R}(e_1, e_4)e_1 = e_3$$
$$\tilde{R}(e_1, e_4)e_4 = -e_2 \tag{4.52}$$

if $f = -(x_1)^2$, where $e_i = \partial/\partial x_i$, $i = 1, \ldots, 4$.

The curvature tensors determined by the equations (4.51) and (4.52) yield two symmetric holonomy systems (V, R, H) and $(\tilde{V}, \tilde{R}, \tilde{H})$ as follows: when

V and $\tilde{V}$ are identified with $\mathbb{R}^4$, H and $\tilde{H}$ are the connected Lie subgroups of $GL(4, \mathbb{R})$ with Lie algebra $\mathfrak{h} = span\{R_{xy}; x, y \in V\}$ and $\tilde{\mathfrak{h}} = span\{\tilde{R}_{xy}; x, y \in \tilde{V}\}$, respectively, and, R and $\tilde{R}$ are the curvature tensors on V and $\tilde{V}$ given by (4.51) and (4.52), respectively. Following [142], let G/H and $\tilde{G}/\tilde{H}$ be the associated symmetric spaces, where G and $\tilde{G}$ are the simply connected Lie groups with Lie algebras $\mathfrak{g}$ and $\tilde{\mathfrak{g}}$, respectively, given by

$$\mathfrak{g} = \mathfrak{h} \oplus V, \qquad \tilde{\mathfrak{g}} = \tilde{\mathfrak{h}} \oplus \tilde{V}$$

where the Lie algebra product is defined by

$$\mathfrak{g}: \begin{cases} [\mathfrak{h}, \mathfrak{h}] = [R_{e_1 e_4}, R_{e_1 e_4}] = 0, \\ [\mathfrak{h}, X] = R_{e_1 e_4}(X), & X \in V, \\ [X, Y] = R_{XY}, & X, Y \in V \end{cases}$$

$$\tilde{\mathfrak{g}}: \begin{cases} [\tilde{\mathfrak{h}}, \tilde{\mathfrak{h}}] = [\tilde{R}_{e_1 e_4}, \tilde{R}_{e_1 e_4}] = 0, \\ [\tilde{\mathfrak{h}}, \tilde{X}] = \tilde{R}_{e_1 e_4}(\tilde{X}), & \tilde{X} \in \tilde{V}, \\ [\tilde{X}, \tilde{Y}] = \tilde{R}_{\tilde{X}\tilde{Y}}, & \tilde{X}, \tilde{Y} \in \tilde{V} \end{cases}$$

Now, if $\{\omega^0, \ldots, \omega^4\}$ denotes a basis for the 1-forms dual to $\{e_1, e_4, R_{e_1 e_4}, e_2, e_3\}$, where $\{e_1, \ldots, e_4\}$ is the canonical basis for $V = \mathbb{R}^4$, then the case (ii) of Theorem 4.2.7 is obtained. Similarly, the case (iii) is obtained by putting $\{\tilde{\omega}^0, \ldots, \tilde{\omega}^4\}$ as the dual basis for $\{e_1, e_4, \tilde{R}_{e_1 e_4}, e_2, e_3\}$. $\qquad\square$

Remark 4.2.6. Among the different possibilities for the Jacobi operators in Theorem 4.2.1, only those corresponding to Type Ia (real space forms and rank-one symmetric spaces) and Type II (cases (ii) and (iii)) may occur as symmetric spaces.

Remark 4.2.7. The Lie groups G and $\tilde{G}$ in Theorem 4.2.7 are 3-step nilpotent Lie groups and they can be realized as matrix groups in the following way. By integrating the structural equations in Theorem 4.2.7-(ii), we have

$$\omega^0 = dx_0, \qquad\qquad \omega^1 = dx_1,$$
$$\omega^2 = dx_2 - x_0 dx_1, \qquad \omega^3 = dx_3 - x_2 dx_1,$$
$$\omega^4 = dx_4 + x_2 dx_0 + \tfrac{1}{2}(x_0)^2 dx_1,$$

where $(x_0, \ldots, x_4)$ are the usual coordinates on $\mathbb{R}^5$. Therefore, the product $*$ of the Lie group G can be described by

$$(a_0, a_1, a_2, a_3, a_4) * (x_0, x_1, x_2, x_3, x_4)$$
$$= (a_0 + x_0, a_1 + x_1, a_2 + x_2 + a_0 x_1,$$
$$a_3 + x_3 + a_2 x_1 + \tfrac{1}{2} a_0 (x_1)^2, a_4 + x_4 - a_2 x_0 - a_0 x_0 x_1 - \tfrac{1}{2}(a_0)^2 x_1),$$

which corresponds to the product of the matrix group given by

$$G = \left\{ \begin{pmatrix} 1 & x_0 & x_2 & x_3 & \frac{1}{2}(x_0)^2 & x_4 \\ 0 & 1 & x_1 & \frac{1}{2}(x_1)^2 & x_0 & -x_0 x_1 \\ 0 & 0 & 1 & x_1 & 0 & -x_0 \\ 0 & 0 & 0 & 1 & 0 & 0 \\ 0 & 0 & 0 & 0 & 1 & -x_1 \\ 0 & 0 & 0 & 0 & 0 & 1 \end{pmatrix} ; x_0, x_1, x_2, x_3, x_4 \in \mathbb{R} \right\}.$$

Similarly, from the structural equations in Theorem 4.2.7-(iii), the Lie group $\tilde{G}$ can be described by the product,

$$(a_0, a_1, a_2, a_3, a_4) \tilde{*} (x_0, x_1, x_2, x_3, x_4)$$
$$= (a_0 + x_0, a_1 + x_1, a_2 + x_2 + a_0 x_1,$$
$$a_3 + x_3 - a_2 x_1 - \tfrac{1}{2} a_0 (x_1)^2, a_4 + x_4 + a_2 x_0 + a_0 x_0 x_1 + \tfrac{1}{2}(a_0)^2 x_1),$$

where $(a_0, \ldots, a_4)$, $(x_0, \ldots, x_4) \in \mathbb{R}^5$, which corresponds to the product of the matrix group given by

$$\tilde{G} = \left\{ \begin{pmatrix} 1 & x_0 & x_2 & x_3 & \frac{1}{2}(x_0)^2 & x_4 \\ 0 & 1 & x_1 & -\frac{1}{2}(x_1)^2 & x_0 & x_0 x_1 \\ 0 & 0 & 1 & -x_1 & 0 & x_0 \\ 0 & 0 & 0 & 1 & 0 & 0 \\ 0 & 0 & 0 & 0 & 1 & x_1 \\ 0 & 0 & 0 & 0 & 0 & 1 \end{pmatrix} ; x_0, x_1, x_2, x_3, x_4 \in \mathbb{R} \right\}.$$

Remark 4.2.8. Note that the symmetric space $\tilde{G}/\tilde{H}$ in Theorem 4.2.7-(iii) corresponds to the example given by Rakić in [125].

A semi-Riemannian manifold (M, g) is said to be *recurrent* if there exists a one-form ω on M such that $\nabla R = \omega \otimes R$. Clearly, any locally symmetric space is recurrent. Moreover, it is interesting to note that there is no scarcity of nonsymmetric recurrent spaces. (See [126] for more information on the geometry of recurrent spaces.)

The result in Theorem 4.2.7 can be partially extended to the broader class of recurrent spaces as follows:

Theorem 4.2.8. [20] *Let (M, g) be a recurrent semi-Riemannian Jordan-Osserman manifold with metric tensor of signature $(2, 2)$. Then it is locally symmetric or Ricci flat of Type II.*

4.2.4 Classification of Semi-Riemannian Jordan-Osserman Manifolds with Metric Tensors of Signature $(2, 2)$

It should be noted here that a complete classification of semi-Riemannian Jordan-Osserman manifolds with metric tensors of signature $(2, 2)$ is not yet known. However, important progresses have been made by Blažić, Bokan and Rakić in [18]. There the different cases appearing in Theorem 4.2.1 are investigated separately. This leads to a complete classification of Type Ia semi-Riemannian Jordan-Osserman manifolds with metric tensors of signature $(2, 2)$ and to the nonexistence of such manifolds of Type Ib. These results are stated as Theorem 4.2.9 and Theorem 4.2.10 below. The results concerning Types II and III are not so conclusive, which we will pay attention to at the end of this subsection.

In what remains of this subsection, we use the following notation. Let $\{E_1, E_2, E_3, E_4\}$ be an orthonormal local frame on (M, g), and put

$$\nabla_{E_i} = \sum \omega_i^s E_s \qquad \text{connection 1-forms}$$

$$\Omega_j^i = \tfrac{1}{2} \sum R_{klj}^i \theta^k \wedge \theta^l \qquad \text{curvature forms}$$

where θ^i are the dual forms to E_i and $R(E_k, E_l)E_j = \sum R_{klj}^i E_i$. Then the Cartan structural equations are given by

$$d\theta^i = -\sum \omega_j^i \wedge \theta^j$$

$$d\omega_j^i = -\sum (\omega_k^i \wedge \omega_j^k + \Omega_j^i)$$

and one has the following symmetry properties,

$$\omega_j^i = -\varepsilon_i \varepsilon_j \omega_i^j \qquad \Omega_j^i = -\varepsilon_i \varepsilon_j \Omega_i^j.$$

Osserman 4-manifolds with diagonalizable Jacobi operators are completely determined as follows.

Theorem 4.2.9. [18] *Let (M, g) be a connected semi-Riemannian manifold with metric tensor of signature $(2, 2)$. If (M, g) is Osserman of Type Ia, then it is locally isometric to,*

(a) a real space form,
(b) a complex space form,
(c) a paracomplex space form.

Proof. As a consequence of Theorem 4.2.7 and the hypothesis above on the diagonalizability of Jacobi operators, it suffices to show that any Type Ia Osserman manifold is locally symmetric. In order to do that, let $R_{jkli;h}$ denote the covariant derivative $(\nabla_{E_h} R)(E_j, E_k, E_l, E_i)$ and show that $R_{jkli;h} = 0$ for all $i, j, k, l, h = 1, \ldots, 4$. Now we proceed as in [18].

Using the expression of the components of the curvature tensor of a Type Ia semi-Riemannian Osserman manifold with metric tensor of signature $(2,2)$ in Theorem 4.2.2, one can compute all curvature 2-forms Ω_i^j as follows:

$$\Omega_3^2 = \Omega_2^3 = y\theta^1 \wedge \theta^4 + \gamma\theta^2 \wedge \theta^3 \qquad \Omega_1^4 = \Omega_4^1 = y\theta^2 \wedge \theta^3 + \gamma\theta^1 \wedge \theta^4$$

$$\Omega_4^2 = \Omega_2^4 = (x+y)\theta^1 \wedge \theta^3 + \beta\theta^2 \wedge \theta^4 \qquad \Omega_3^1 = \Omega_1^3 = (x+y)\theta^2 \wedge \theta^4 + \beta\theta^1 \wedge \theta^3$$

$$\Omega_4^3 = -\Omega_3^4 = -x\theta^1 \wedge \theta^2 + \alpha\theta^3 \wedge \theta^4 \qquad \Omega_1^2 = -\Omega_2^1 = -x\theta^3 \wedge \theta^4 + \alpha\theta^1\theta^2.$$

$$(4.53)$$

First observe that the eigenvalues α, β and γ of Jacobi operators cannot be all different. In fact, by using the differential,

$$d\Omega_i^j = \sum_s (\Omega_s^j \wedge \omega_i^s - \omega_s^j \wedge \Omega_i^s)$$

of the structural equations, one obtains the following:

$$(\alpha - \gamma)B \wedge (\theta^1 \wedge \theta^4 - \theta^2 \wedge \theta^3) + (\beta - \alpha)C \wedge (\theta^1 \wedge \theta^3 + \theta^2 \wedge \theta^4) = 0$$

$$(\beta - \gamma)A \wedge (\theta^2 \wedge \theta^3 - \theta^1 \wedge \theta^4) + (\beta - \alpha)C \wedge (\theta^1 \wedge \theta^2 - \theta^3 \wedge \theta^4) = 0 \quad (4.54)$$

$$(\beta - \gamma)A \wedge (\theta^1 \wedge \theta^3 + \theta^2 \wedge \theta^4) + (\gamma - \alpha)B \wedge (\theta^1 \wedge \theta^2 + \theta^3 \wedge \theta^4) = 0$$

where

$$A = \omega_1^2 - \omega_3^4 = A_1\theta^1 + A_2\theta^2 + A_3\theta^3 + A_4\theta^4$$

$$B = \omega_1^3 + \omega_2^4 = B_2\theta^1 + B_2\theta^2 + B_3\theta^3 + B_4\theta^4 \qquad (4.55)$$

$$C = \omega_1^4 - \omega_2^3 = C_1\theta^1 + C_2\theta^2 + C_3\theta^3 + C_4\theta^4.$$

Now, if α, β, γ are all distinct, one can see that the equations (4.54) form a linear system and by solving it, one obtains

$$B_1 = sA_4, \quad B_2 = -sA_3, \quad B_3 = -sA_2, \quad B_4 = sA_1,$$

$$C_1 = tA_3, \quad C_2 = tA_4, \quad C_3 = tA_1, \quad C_4 = tA_2, \qquad (4.56)$$

where

$$s = \frac{\beta - \gamma}{\alpha - \gamma} \neq 0, \qquad t = \frac{\beta - \gamma}{\alpha - \beta} \neq 0.$$

Now, we introduce 1-forms φ^2, φ^3 and φ^4 as follows:

$$A = \omega_1^2 - \omega_3^4 = A_1\theta^1 + A_2\theta^2 + A_3\theta^3 + A_4\theta^4 = \varphi^2,$$

$$B = \omega_1^3 + \omega_2^4 = s(AA_4\theta^1 - A_3\theta^2 - A_2\theta^3 + A_1\theta^4) = s\varphi^3, \qquad (4.57)$$

$$C = \omega_1^4 - \omega_2^3 = t(A_3\theta^1 + A_4\theta^2 + A_1\theta^3 + A_2\theta^4) = t\varphi^4.$$

Using the Cartan structural equations we have

$$d\varphi^2 = \Omega_1^2 - \Omega_3^4 - st\varphi^3 \wedge \varphi^4,$$
$$d\varphi^3 = \Omega_1^4 + \Omega_2^4 - t\varphi^2 \wedge \varphi^4, \qquad (4.58)$$
$$d\varphi^4 = \Omega_1^4 - \Omega_2^3 - s\varphi^3 \wedge \varphi^2.$$

Then, by computing the covariant derivatives of the forms A, B and C, one obtains

$$A_{2;1} - A_{1;2} = \alpha - x - st(A_3^2 + A_4^2)$$
$$A_{4;3} - A_{3;4} = \alpha - x + st(A_1^2 + A_2^2)$$
$$-s(A_{2;1} + A_{4;3}) = \beta + x + y - t(A_1^2 - A_3^2)$$
$$s(A_{1;2} + A_{3;4}) = \beta + x + y - t(A_2^2 - A_4^2) \qquad (4.59)$$
$$t(A_{2;1} - A_{3;4}) = \gamma - y - s(A_4^2 - A_1^2)$$
$$t(A_{1;2} - A_{4;3}) = y - \gamma - s(A_2^2 - A_3^2).$$

Next, let

$$\varphi^1 = A_2\theta^1 - A_1\theta^2 - A_4\theta^3 + A_3\theta^4 \qquad (4.60)$$

and note that $\varphi^1, \varphi^2, \varphi^3, \varphi^4$ form an orthogonal basis for T_p^*M. Now, computing the divergence of φ^1,

$$div\varphi^1 = -A_{2;1} + A_{1;2} - A_{4;3} + A_{3;4},$$
$$div\varphi^1 = 2(x - \alpha) + st\mathbb{I},$$
$$sdiv\varphi^1 = 2(\beta + x + y) + t\mathbb{I}, \qquad (4.61)$$
$$tdiv\varphi^1 = 2(y - \gamma) + s\mathbb{I},$$

where $\mathbb{I} = -A_1^2 - A_2^2 + A_3^2 + A_4^2 = ||A||^2$. Now, from the last three equations in (4.61), one obtains

$$3div\varphi^1 = \frac{\mathbb{I}}{(\alpha - \gamma)(\alpha - \beta)}[(\beta - \gamma)^2 + (\alpha - \gamma)^2 + (\alpha - \beta)^2]. \qquad (4.62)$$

Hence, we have two possibilities: if $\mathbb{I} = 0$, then it follows from (4.61) that

$$\alpha - x = \frac{\beta + x + y}{-s} = \frac{\gamma - y}{t}$$

and thus $\alpha = \beta = \gamma = 0$. Therefore, let us assume $\mathbb{I} \neq 0$. Let $\{\eta_i\}$ denote the basis for T_pM dual to $\{\varphi^i\}$ and put

$$div\varphi^1 = \Upsilon_1^2(\eta_2) + \Upsilon_1^3(\eta_3) + \Upsilon_1^4(\eta_4), \qquad (4.63)$$

where Υ_j^i are the connection forms with respect to the basis $\{\eta_i\}$. Then

$$\varphi^1 \wedge \varphi^2 - \varphi^3 \wedge \varphi^4 = -\mathbb{I}(\theta^1 \wedge \theta^2 + \theta^3 \wedge \theta^4)$$
$$\varphi^1 \wedge \varphi^3 - \varphi^2 \wedge \varphi^4 = \mathbb{I}(\theta^1 \wedge \theta^3 + \theta^2 \wedge \theta^4) \qquad (4.64)$$
$$\varphi^1 \wedge \varphi^4 + \varphi^2 \wedge \varphi^3 = -\mathbb{I}(\theta^1 \wedge \theta^4 - \theta^2 \wedge \theta^3).$$

Now, after some computations, it follows that

$$\Upsilon_1^2(\eta_2) = \frac{x - \alpha}{\mathbb{I}}, \quad \Upsilon_1^3(\eta_3) = \frac{\beta + x + y}{s\mathbb{I}}, \quad \Upsilon_1^4(\eta_4) = \frac{y - \gamma}{t\mathbb{I}}. \qquad (4.65)$$

Finally, a direct computation from (4.63) and (4.65) leads to $div\varphi^1 = 0$, which is a contradiction with $\mathbb{I} \neq 0$.

Therefore, from (4.54), we have:

$$\begin{aligned}
&\text{If } \alpha = \beta \neq \gamma \quad \text{then } \omega_1^3 + \omega_2^4 = 0 \text{ and } \omega_1^2 - \omega_3^4 = 0. \\
&\text{If } \alpha = \gamma \neq \beta \quad \text{then } \omega_2^3 - \omega_1^4 = 0 \text{ and } \omega_1^2 - \omega_3^4 = 0. \qquad (4.66) \\
&\text{If } \beta = \gamma \neq \alpha \quad \text{then } \omega_2^3 - \omega_1^4 = 0 \text{ and } \omega_1^3 + \omega_2^4 = 0.
\end{aligned}$$

Now, from the symmetry properties of the connection forms and the expression of the components of the curvature tensor in Theorem 4.2.2, we have

$$R_{ijkl;h} = 0 \qquad \text{and} \qquad R_{ijji;h} = 0 \qquad (4.67)$$

provided that i, j, k, l are different. Moreover, for the remaining components of the covariant derivative of R, we have

$$\begin{aligned}
&\sum_h R_{1242;h}\theta^h = -\sum_h R_{2131;h}\theta^h = (\beta - \alpha)(\omega_2^3 - \omega_1^4) \\
&\sum_h R_{2132;h}\theta^h = \sum_h R_{1241;h}\theta^h = (\gamma - \alpha)(\omega_2^4 + \omega_1^3) \qquad (4.68) \\
&-\sum_h R_{3123;h}\theta^h = \sum_h R_{1341;h}\theta^h = (\beta - \gamma)(\omega_1^2 - \omega_3^4).
\end{aligned}$$

Now, it follows from (4.66) and (4.68) that

$$R_{1242;h} = R_{2131;h} = R_{2132;h} = R_{1241;h} = R_{3123;h} = R_{1341;h} = 0.$$

Putting this together with (4.67), we conclude that (M, g) is locally symmetric and the result follows from Theorem 4.2.7. $\qquad \square$

Remark 4.2.9. The proof of the Riemannian version of Theorem 4.2.9 (cf. Theorem 2.1.7) follows the same steps as in the proof above. However, note that only cases (a) and (b) are available in the Riemannian setting.

The situation when the Jacobi operators are nondiagonalizable is as follows:

Theorem 4.2.10. *Let (M, g) be a connected semi-Riemannian manifold with metric tensor of signature $(2, 2)$. Then it cannot be Osserman of Type Ib. Moreover, if (M, g) is Osserman of Type II or Type III, then the Jacobi operators have only one eigenvalue.*

The structures of Type II and III Osserman manifolds have been investigated in [22], where the Authors prove the following:

Theorem 4.2.11. [22] *A semi-Riemannian Osserman manifold with metric tensor of signature* $(2, 2)$ *and Types II or III is locally foliated by totally geodesic, flat, isotropic two-dimensional submanifolds.*

Moreover, any Type II or III Osserman manifold can be locally described as follows:

Theorem 4.2.12. [22] *Let* (M, g) *be a semi-Riemannian Osserman manifold with metric tensor of signature* $(2, 2)$. *If Jacobi operators are of either Type II or III, then there exists a coordinate system* $(U, (u_1, u_2, u_3, u_4))$ *such that the metric tensor is given by*

$$g = \begin{bmatrix} 0 & 0 & g_{13} & g_{14} \\ 0 & 0 & g_{23} & g_{24} \\ g_{13} & g_{23} & 0 & g_{34} \\ g_{14} & g_{24} & g_{34} & 0 \end{bmatrix} \quad \text{with} \quad \det \begin{pmatrix} g_{13} & g_{14} \\ g_{23} & g_{24} \end{pmatrix} = 1.$$

5. Semi-Riemannian Osserman Manifolds

5.1 Nonsymmetric Semi-Riemannian Osserman Manifolds of Arbitrary Signature

A positive solution to the Osserman problem in the Riemannian setting was provided under some hypothesis on the dimension of the manifold or on the existence of special additional structures on the manifold. In this subsection we will point out the nonexistence of analogous results in the general semi-Riemannian setting. First of all we know the *existence of nonsymmetric semi-Riemannian Osserman manifolds with metric tensors of arbitrary signature* (ν, η), ν, $\eta > 1$. For, let (M_1, g_1) and (M_2, g_2) be semi-Riemannian manifolds of dimensions n_1 and n_2, respectively. The product $M = M_1 \times M_2$ furnished with the product metric tensor $g = g_1 \oplus g_2$ is also a semi-Riemannian manifold. Moreover, for each vector field $X = (X_1, X_2)$ tangent to M, where X_1 and X_2 are vector fields on M_1 and M_2, respectively, the characteristic polynomial of the operator $R_X = R(\cdot, X)X$ on TM satisfies

$$p_\lambda(R_X) = \det(R_X - \lambda I_{n_1+n_2}) = \det(R_{X_1}^{(1)} - \lambda I_{n_1})\det(R_{X_2}^{(2)} - \lambda I_{n_2})$$

where $R^{(1)}$ and $R^{(2)}$ are the curvature tensors of (M_1, g_1) and (M_2, g_2), respectively, and $R_X^{(1)} = R^{(1)}(\cdot, X_1)X_1$ on TM_1 and $R_X^{(2)} = R^{(2)}(\cdot, X_2)X_2$ on TM_2.

Now, it follows that the product manifold (M, g) is Osserman if both factors (M_1, g_1) and (M_2, g_2) are Osserman with vanishing eigenvalues of their Jacobi operators. (Note that, in the Riemannian case, a pointwise Osserman manifold is flat if it is locally reducible [9, Lemma 2.2].)

Therefore, if $(N, g_{(f_1, f_2)})$ is one of the examples constructed in Section 4.1, then the product manifold $\mathbb{R}^n \times N$ endowed with the product metric tensor $g = g_{(\nu-2, \eta-2)} \oplus g_{(f_1, f_2)}$ is a semi-Riemannian manifold with metric tensor of signature (ν, η) that is Osserman but not locally symmetric, where $g_{(\nu-2, \eta-2)}$ denotes the semi-Euclidean metric tensor of signature $(\nu-2, \eta-2)$, $\nu, \eta \geq 2$.

Remark 5.1.1. Note that, at each point of $(\mathbb{R}^n \times N,\ g)$ there exist unit vectors whose Jacobi operators have different minimal polynomials. Indeed, for $x = (x_1, 0)$, the associated Jacobi operator vanishes identically and thus the minimal polynomial of R_x becomes $m_\lambda(R_x) = \lambda$. However, since R_x is nonzero for $x = (0, x_2)$, where $(N, g_{(f_1, f_2)})$ is choosen as in Remark 4.1.2, its minimal polynomial is $m_\lambda(R_x) = \lambda^s$, $s = 1, 2, 3$, depending on the point and the metric tensor $g_{(f_1, f_2)}$ defined on N. This shows that the Jacobi operators may be diagonalizable for some directions and may not be diagonalizable for other directions at a point.

Remark 5.1.2. For any locally symmetric semi-Riemannian Osserman manifold with $p_\lambda(R_x) = \lambda^4$ and $m_\lambda(R_x) = \lambda^2$ (cf. Theorem 4.2.7), the product manifolds constructed above are also locally symmetric. Note that, even in this case the minimal polynomial of the Jacobi operator has nonconstant roots, counting multiplicities, at each point. This shows the existence of *symmetric Osserman manifolds that are not Jordan-Osserman.*

5.1.1 Indefinite Kähler Osserman Manifolds

In Chapter 1, it is shown that the assumption of additional structures on a Riemannian Osserman manifold led to affirmative conclusions in the solution to the Osserman problem. Among such structures, the Kähler ones are well-understood (cf. Section 2.2.1.) At this point, it is important to recall the special role played by the holomorphic sectional curvature in Lemma 2.2.1. Hence one may expect to obtain some similar results in the semi-Riemannian case. For that, a closer examination of the holomorphic sectional curvature is needed.

The holomorphic sectional curvature is a real-valued function defined on the unit sphere bundle of an almost Hermitian manifold with Riemannian metric tensor and hence it is bounded at each point. However, for indefinite metric tensors, such a statement is no longer true, that is, the holomorphic sectional curvature of an indefinite almost Hermitian manifold is bounded (from above and below) at a point if and only if it is constant at that point [4], [27]. Therefore, one may expect to obtain affirmative conclusions in the study of the Osserman problem by imposing weak boundedness conditions on the holomorphic sectional curvature of an indefinite Kähler manifold. For example, one may restrict the condition of boundedness (from above and below) on curvature of holomorphic planes of signature $(++)$ or $(--)$. However this also yields constant holomorphic sectional curvature [27], [100]. Therefore a strictly weaker assumption on the holomorphic sectional curvature of a Kähler manifold is to assume that the curvature is bounded from below (or from above) on the holomorphic planes of signature $(+, +)$, and from above (or from below) on the holomorphic planes of signature $(-, -)$. This condition is equivalent to the vanishing Bochner tensor of the manifod [29], [100]. Now, since any Osserman manifold is Einstein, it immediately follows again that the indefinite Kähler Osserman manifolds whose curvature is bounded from below (or from above) on the holomorphic planes of signature $(+, +)$, and from above (or from below) on the holomorphic planes of signature $(-,-)$, are of constant holomorphic sectional curvature.

A further weaker condition on the holomorphic sectional curvature is to assume that the curvature is bounded from either above or below on non-degenerate holomorphic planes. This condition is strictly weaker than the constant holomorphic sectional curvature (cf. [4]) and we show here that this kind of boundedness conditions does not yield an affirmative answer to the Osserman problem in semi-Riemannian geometry.

Let us consider 4-dimensional space $\mathbb{R}^4$ with coordinates (x_1, x_2, x_3, x_4). Let $\phi = (\phi_{ij})$ be a symmetric $(0, 2)$-tensor field on $\mathbb{R}^2$ and consider the metric tensor g_ϕ on $\mathbb{R}^4$, defined by

$$g_\phi = dx_1 \otimes dx_3 + dx_2 \otimes dx_4 + \sum_{i,j=1}^{2} \phi_{ij} dx_i \otimes dx_j. \qquad (5.1)$$

The Levi-Civita connection of $(\mathbb{R}^4, g_\phi)$ is given by

$$\nabla_{\frac{\partial}{\partial x_1}} \frac{\partial}{\partial x_1} = \frac{1}{2} \frac{\partial \phi_{11}}{\partial x_1} \frac{\partial}{\partial x_3} + [-\frac{1}{2} \frac{\partial \phi_{11}}{\partial x_2} + \frac{\partial \phi_{12}}{\partial x_1}] \frac{\partial}{\partial x_4},$$

$$\nabla_{\frac{\partial}{\partial x_1}} \frac{\partial}{\partial x_2} = \frac{1}{2} \frac{\partial \phi_{11}}{\partial x_2} \frac{\partial}{\partial x_3} + \frac{1}{2} \frac{\partial \phi_{22}}{\partial x_1} \frac{\partial}{\partial x_4}, \qquad (5.2)$$

$$\nabla_{\frac{\partial}{\partial x_2}} \frac{\partial}{\partial x_2} = [-\frac{1}{2} \frac{\partial \phi_{22}}{\partial x_1} + \frac{\partial \phi_{12}}{\partial x_2}] \frac{\partial}{\partial x_3} + \frac{1}{2} \frac{\partial \phi_{22}}{\partial x_2} \frac{\partial}{\partial x_4}.$$

Hence the nonvanishing components of the curvature tensor are

$$R(\frac{\partial}{\partial x_1}, \frac{\partial}{\partial x_2}) \frac{\partial}{\partial x_1} = [\frac{1}{2} \frac{\partial^2 \phi_{11}}{\partial x_2 \partial x_2} + \frac{1}{2} \frac{\partial^2 \phi_{22}}{\partial x_1 \partial x_1} - \frac{\partial^2 \phi_{12}}{\partial x_1 \partial x_2}] \frac{\partial}{\partial x_4},$$

$$R(\frac{\partial}{\partial x_1}, \frac{\partial}{\partial x_2}) \frac{\partial}{\partial x_2} = -[\frac{1}{2} \frac{\partial^2 \phi_{11}}{\partial x_2 \partial x_2} + \frac{1}{2} \frac{\partial^2 \phi_{22}}{\partial x_1 \partial x_1} - \frac{\partial^2 \phi_{12}}{\partial x_1 \partial x_2}] \frac{\partial}{\partial x_3}. \qquad (5.3)$$

Thus, we have

Theorem 5.1.1. $(\mathbb{R}^4, g_\phi)$ *is a semi-Riemannian Osserman manifold which is locally symmetric if and only if*

$$\frac{1}{2} \frac{\partial^2 \phi_{11}}{\partial x_2 \partial x_2} + \frac{1}{2} \frac{\partial^2 \phi_{22}}{\partial x_1 \partial x_1} - \frac{\partial^2 \phi_{12}}{\partial x_1 \partial x_2} \qquad (5.4)$$

is a constant function, where ϕ_{ij}, $i, j = 1, 2$, *are the components of a symmetric* $(0, 2)$-*tensor field* ϕ *on* $\mathbb{R}^2$.

Moreover, the characteristic polynomial of the Jacobi operators is always $p_\lambda(R_x) = \lambda^4$, *while the minimal polynomial* $m_\lambda(R_x)$ *is given by*

(i) $(\mathbb{R}^4, g_\phi)$ has zero sectional curvature (that is, $m_\lambda(R_x) = \lambda$) at any point where

$$\frac{1}{2}\frac{\partial^2 \phi_{11}}{\partial x_2 \partial x_2} + \frac{1}{2}\frac{\partial^2 \phi_{22}}{\partial x_1 \partial x_1} - \frac{\partial^2 \phi_{12}}{\partial x_1 \partial x_2}$$

vanishes,

(ii) the minimal polynomial is $m_\lambda(R_x) = \lambda^2$ at those points where

$$\frac{1}{2}\frac{\partial^2 \phi_{11}}{\partial x_2 \partial x_2} + \frac{1}{2}\frac{\partial^2 \phi_{22}}{\partial x_1 \partial x_1} - \frac{\partial^2 \phi_{12}}{\partial x_1 \partial x_2}$$

is different from zero.

Proof. The proof follows from (5.2) and (5.3) by proceeding analogously as in the proofs of Theorems 4.1.1 and 4.1.2. □

Now, let J denote the complex structure on $\mathbb{R}^4$ defined by $J\frac{\partial}{\partial x_1} = \frac{\partial}{\partial x_2}$ and $J\frac{\partial}{\partial x_3} = \frac{\partial}{\partial x_4}$. A symmetric $(0,2)$-tensor field ϕ on $(\mathbb{R}^2, J)$ is called Hermitian if it satisfies $\phi(JX, JY) = \phi(X, Y)$ for all $X, Y \in \Gamma\mathbb{R}^2$, where $J\frac{\partial}{\partial x_1} = \frac{\partial}{\partial x_2}$ denotes the usual complex structure on $\mathbb{R}^2$. (Equivalently, $\phi_{11} = \phi_{22}$ and $\phi_{12} = \phi_{21} = 0$.) Note from (5.1) that g_ϕ is an indefinite almost Hermitian metric tensor on $(\mathbb{R}^4, J)$ if and only if ϕ is Hermitian. Moreover, we have:

Theorem 5.1.2. Let ϕ be a Hermitian $(0,2)$-tensor field on $(\mathbb{R}^2, J)$. Then $(\mathbb{R}^4, g_\phi, J)$ is an indefinite Kähler Osserman manifold and moreover, the holomorphic sectional curvature of $(\mathbb{R}^4, g_\phi, J)$ has the same sign of $\Delta\phi_{11}$, where Δ denotes the Euclidean Laplacian on $\mathbb{R}^2$.

Proof. Since ϕ is Hermitian, $\phi_{11} = \phi_{22}$ and $\phi_{12} = \phi_{21} = 0$. Then it follows from (5.2) after a straightforward computation that $(\mathbb{R}^4, g_\phi, J)$ is an indefinite Kähler manifold. Moreover, the expression (5.3) of the curvature tensor becomes

$$R\left(\frac{\partial}{\partial x_1}, \frac{\partial}{\partial x_2}\right)\frac{\partial}{\partial x_1} = \frac{1}{2}\Delta\phi_{11}\frac{\partial}{\partial x_4},$$

$$R\left(\frac{\partial}{\partial x_1}, \frac{\partial}{\partial x_2}\right)\frac{\partial}{\partial x_2} = -\frac{1}{2}\Delta\phi_{11}\frac{\partial}{\partial x_3}$$

where $\Delta = \frac{\partial^2}{\partial x_1 \partial x_1} + \frac{\partial^2}{\partial x_2 \partial x_2}$ is the Euclidean Laplacian on $\mathbb{R}^2$. Thus, the curvature of a nondegenerate holomorphic plane $span\{X, JX\}$ is given by

$$H(span\{X, JX\}) = \frac{g_\phi(R(X, JX)X, JX)}{g_\phi(X, X)^2} = \frac{1}{2}\left(\frac{\alpha_1^2 + \alpha_2^2}{g_\phi(X, X)}\right)^2 \Delta\phi_{11},$$

where $X = \sum_i \alpha_i \frac{\partial}{\partial x_i}$. This shows that the holomorphic sectional curvature has the sign of $\Delta\phi_{11}$. □

Corollary 5.1.1. *If $(\mathbb{R}^4, g_\phi, J)$ is a locally symmetric indefinite Kähler Osserman manifold, then the holomorphic sectional curvature of $(\mathbb{R}^4, g_\phi, J)$ is either nonpositive or nonnegative, but not constant unless it is flat.*

Proof. Note that, if ϕ is a Hermitian $(0,2)$-tensor field on $\mathbb{R}^2$, then (5.4) is reduced to $\Delta\phi_{11}$, and hence the result follows. □

Remark 5.1.3. Note that the product of the manifolds $(\mathbb{R}^4, g_\phi, J)$ and semi-Euclidean spaces allows us to construct indefinite Kähler manifolds of any signature $(2\nu, 2\eta)$, $\nu, \eta \geq 1$, with nonnegative or nonpositive holomorphic sectional curvature that are Osserman, but not locally symmetric.

Remark 5.1.4. Analogous results to Theorems 5.1.2 and 5.1.1 and to Corollary 5.1.1 in the framework of para-Hermitian geometry can be found in [31].

5.1.2 Semi-Riemannian Osserman Metric Tensors on Tangent Bundles

In this subsection we construct a broad family of semi-Riemannian Osserman metric tensors of signature (n, n) on the tangent bundle of certain n-dimensional Riemannian manifolds (cf. Theorem 5.1.3 below.) It should be noted that the metric tensors (5.1) are a particular case of the more general situation described below.

First, we recall some basic facts from the geometry of tangent bundles (see [145] for more details.) The tangent bundle of a manifold M consists of pairs (p, x), where $p \in M$ and $x \in T_p M$. Let $\pi : TM \to M$ defined by $\pi(p, x) = p$ be the natural projection of TM onto M. In order to describe the geometry of TM, we make a natural *lift* construction as follows.

For a function $f : M \to \mathbb{R}$, define its vertical lift f^V by $f^V = f \circ \pi$. Note that any 1-form ω on M can be viewed as a function $i\omega$ on TM defined by $i\omega(p, z) = \omega_p(z)$. Therefore, for any given function on M, let $i(df)$ be the function on TM associated with df. It is important to note here that such kind of functions on TM have special significance since two vector fields $\tilde{X}, \tilde{Y}$ on TM satisfying $\tilde{X}(i(df)) = \tilde{Y}(i(df))$ for all functions f on M are necessarily the same. A vector field $\tilde{X}$ on TM is said to be vertical if $\tilde{X}(f^V) = 0$ for all functions f on M. Now, for a given vector field X on M, define the vertical lift X^V of X to TM by $X^V(i\omega) = \omega(X)^V$, where ω is an arbitrary 1-form on M. (Note that X^V is a vertical vector field on TM.) In an analogous way, a 1-form $\tilde{\omega}$ on TM is said to be vertical if $\tilde{\omega}(X^V) = 0$ for all vector fields X on M. For a given function f on M, the vertical lift $(df)^V$ of df to TM is defined by $(df)^V = d(f^V)$ and moreover, if $\omega = \sum \omega_i dx_i$ is a 1-form on M, then its vertical lift to TM is given by $\omega^V = \sum \omega_i^V (dx_i)^V$, where (x_i) is a coordinate chart on M. This notion can now inductively be extended to tensor fields by using the following equations:

$$(P \otimes Q)^V = P^V \otimes Q^V, \qquad (P + Q)^V = P^V + Q^V.$$

If f is a function on M, define the complete lift f^C of f to TM by $f^C = i(df)$. Note that two vector fields $\tilde{X}, \tilde{Y}$ on TM such that $\tilde{X}(f^C) = \tilde{Y}(f^C)$ for all functions f on M are necessarily the same. Hence, define the complete lift of a vector field X on M to be the vector field X^C on TM given by $X^C(f^C) = (Xf)^C$. Also note that 1-forms on TM are characterized by their values on complete lifts of vector fields on M and thus, define the complete lift of a 1-form ω by $\omega^C(X^C) = (\omega(X))^C$. Now, the complete lift procedure can be extended to tensor fields by using the equations

$$(P \otimes Q)^C = P^C \otimes Q^V + P^V \otimes Q^C, \qquad (P + Q)^C = P^C + Q^C.$$

A local coordinate description of the above lifts can be given as follows. Let $(U, (x_1, \ldots, x_n))$ be a coordinate chart on M. Then, it induces local coordinates $(x_1, \ldots, x_n, x_{\bar{1}}, \ldots, x_{\bar{n}})$ on $\pi^{-1}(U)$, where $x_{\bar{1}}, \ldots, x_{\bar{n}}$ represent the components of vector fields on M with respect to local frame $\{\frac{\partial}{\partial x_1}, \ldots, \frac{\partial}{\partial x_n}\}$ in TM. In the following we use the notation $\bar{i} = i + n$ for all $i = 1, \ldots, n$.

Now, if $X = X^i \frac{\partial}{\partial x_i}$ is a vector field on M, one has

$$X^V = (X^i \frac{\partial}{\partial x_i})^V = X^i \frac{\partial}{\partial x_{\bar{i}}}$$

$$X^C = (X^i \frac{\partial}{\partial x_i})^C = (X^i \circ \pi) \frac{\partial}{\partial x_i} + (x_{\bar{k}} \frac{\partial X^i}{\partial x_k}) \frac{\partial}{\partial x_{\bar{i}}}.$$

Moreover, if ϕ is a $(0, 2)$-tensor field on M with components (ϕ_{ij}), then

$$\phi^V = \begin{pmatrix} \phi_{ij} & 0 \\ 0 & 0 \end{pmatrix},$$

$$\phi^C = \begin{pmatrix} (x_{\bar{k}} \frac{\partial \phi_{ij}}{\partial x_k} & \phi_{ij} \\ \phi_{ij} & 0 \end{pmatrix}$$

with respect to $(x_1, \ldots, x_n, x_{\bar{1}}, \ldots, x_{\bar{n}})$.

Now, let (M, g) be a Riemannian manifold. Then there is a natural induced semi-Riemannian metric tensor g^C on TM, called the *complete lift of g*. It follows immediately from the local expressions above that g^C is a semi-Riemannian metric tensor of signature (n, n) on TM. Although the projection $\pi : (TM, g^C) \to (M, g)$ is not a semi-Riemannian submersion, g^C reflects some interesting properties of (M, g) in a nice way [37]. As a generalization of the complete lift metric tensor, first we introduce a family of metric tensors on TM which are of interest to us. Let ϕ be a symmetric $(0, 2)$-tensor field on (M, g), and define on TM the *deformed complete lift metric tensor g_ϕ* of the Riemannian metric tensor g [115] by

$$g_\phi = g^C + \phi^V. \tag{5.5}$$

With respect to the induced coordinates on TM, we have the following local expression for g_ϕ:

$$g_\phi = \begin{pmatrix} \phi_{ij} + \frac{\partial g_{ij}}{\partial x_k} x_{\bar{k}} & g_{ij} \\ g_{ij} & 0 \end{pmatrix}. \tag{5.6}$$

Thus, we have

Theorem 5.1.3. *Let (M, g) be a Riemannian manifold and ϕ be a symmetric $(0, 2)$-tensor field on M. Then, the tangent bundle TM equipped with the deformed complete lift metric tensor $g_\phi = g^C + \phi^V$ is a semi-Riemannian Osserman manifold with indefinite metric tensor if and only if (M, g) is Osserman with zero eigenvalues of the Jacobi operators.*

Proof. Let $(U, (x_i))$ be a coordinate chart on M, $(\pi^{-1}(U), (x_i, x_{\bar{i}}))$ be the induced chart on TM and $\tilde{X} = \sum_{i=1}^{n} \{\alpha_i \frac{\partial}{\partial x_i} + \alpha_{\bar{i}} \frac{\partial}{\partial x_{\bar{i}}}\}$ be a vector field on TM, where $\{\frac{\partial}{\partial x_i}, \frac{\partial}{\partial x_{\bar{i}}}\}$ are the coordinate vector fields on TM. By a straightforward computation, we obtain the components of the curvature tensor $\tilde{R}$ of (TM, g_ϕ) as follows:

$$\tilde{R}^l_{ij\bar{k}} = \tilde{R}^l_{i\bar{j}k} = \tilde{R}^l_{\bar{i}jk} = \tilde{R}^{\bar{l}}_{ij\bar{k}} = \tilde{R}^{\bar{l}}_{\bar{i}j\bar{k}} = 0,$$

$$\tilde{R}^l_{ijk} = \tilde{R}^{\bar{l}}_{ij\bar{k}} = \tilde{R}^{\bar{l}}_{i\bar{j}k} = R^l_{ijk},$$

$$\tilde{R}^{\bar{l}}_{i\bar{j}k} = \tilde{R}^{\bar{l}}_{\bar{i}j\bar{k}} = \tilde{R}^{\bar{l}}_{\bar{i}\bar{j}k} = 0,$$

where R^l_{ijk} are the components of the curvature tensor of (M, g) with respect to the coordinate system $(x_1, \ldots, x_n)$. Therefore, the matrix form of the operator $\tilde{R}_{\tilde{X}} = \tilde{R}(\cdot, \tilde{X})\tilde{X}$ on TTM with respect to $(x_i, x_{\bar{i}})$ becomes

$$\tilde{R}_{\tilde{X}} = \begin{pmatrix} [R_X] & 0 \\ * & [R_X] \end{pmatrix}, \tag{5.7}$$

where $X = \sum_{i=1}^{n} \alpha_i \frac{\partial}{\partial x_i}$ and $[R_X]$ is the matrix of the operator $R_X = R(\cdot, X)X$ on TM with respect to $\{\frac{\partial}{\partial x_1}, \ldots, \frac{\partial}{\partial x_n}\}$.

Now, let $x = \sum_{i=1}^{n} \alpha_i \frac{\partial}{\partial x_i}$ be a nonnull vector on $T_p M$ and let $\tilde{x} \in T_{\tilde{p}} TM$, where $\tilde{p} = (p, 0)$, be $\tilde{x} = \sum_{i=1}^{n} \{\alpha_i \frac{\partial}{\partial x_i} + \alpha_{\bar{i}} \frac{\partial}{\partial x_{\bar{i}}}\}$. Here $\{\alpha_{\bar{i}}\}$ are defined as follows. Choose $k \in \{1, \ldots, n\}$ in a way that $\sum_{i=1}^{n} \alpha_i g_{ik} \neq 0$ (note that this is always possible since x is a nonnull vector) and define

$$
\alpha_{\tilde{t}} = \begin{cases} \dfrac{1 - \sum_{i,j=1}^{n} \alpha_i \alpha_j \phi_{ij}}{2 \sum_{i=1}^{n} \alpha_i g_{ik}} & \text{if } t = k \\ \\ 0 & \text{if } t \neq k. \end{cases}
$$

Then, since

$$
g_\phi(\tilde{x}, \tilde{x}) = \sum_{i,j,r=1}^{n} \alpha_i \alpha_j x_{\tilde{r}} \frac{\partial g_{ij}}{\partial x_r} + \sum_{i,j=1}^{n} \alpha_i \alpha_j \phi_{ij} + 2 \sum_{i,j=1}^{n} \alpha_i \alpha_{\tilde{j}} g_{ij},
$$

$\tilde{x}$ is a unit vector at the zero section of TM.

Now, if (TM, g_ϕ) is assumed to be Osserman then it follows from (5.7) that (M, g) is also Osserman. Moreover, since the (complex) eigenvalues of the Jacobi operators R_x are independent of $x \in S(M)$, they are necessarily vanishing. Conversely, if (M, g) is assumed to be Osserman with vanishing eigenvalues of Jacobi operators, then it follows from (5.7) that the characteristic polynomial $p_\lambda(\tilde{R}_{\tilde{x}}) = \det(\tilde{R}_{\tilde{x}} - \lambda id_{2n})$ of the Jacobi operator $\tilde{R}_{\tilde{x}}$ satisfies $p_\lambda(\tilde{R}_{\tilde{x}}) = \lambda^{2n}$ for any vector $\tilde{x}$ tangent to TM. This shows that (TM, g_ϕ) is Osserman.   □

Remark 5.1.5. A Riemannian or Lorentzian Osserman manifold with zero eigenvalues of the Jacobi operators is necessarily flat. However, this is no longer true for the other signatures (cf. §4.1 and §5.1.1.) Also note that, if (M, g) is a flat manifold, then (TM, g_ϕ) is a semi-Riemannian Osserman manifold which is not necessarily flat (just considering ϕ to be any conformal change of the metric tensor g.)

5.2 Special Semi-Riemannian Osserman Manifolds

The purpose of this section is to study semi-Riemannian Osserman manifolds (M, g) of higher dimensions under the assumption of the diagonalizability of Jacobi operators. It is clear that, if the Jacobi operator with respect to each $x \in S(M)$ is diagonalizable with a single eigenvalue (which changes sign from spacelike to timelike vectors), then the manifold is a real space form. (Note that all examples constructed in Sections 4.1 and 5.1 have a single zero eigenvalue for the Jacobi operators. However they are nonflat since the Jacobi operators are not diagonalizable.) Therefore, we devote our attention to the first nontrivial case of the Jacobi operators having two distinct eigenvalues.

(I) *For every $x \in S_p(M)$ at each $p \in M$, the Jacobi operator R_x is diagonalizable with exactly two distinct eigenvalues $\varepsilon_x \lambda$ and $\varepsilon_x \mu$, where $\varepsilon_x = g(x, x)$ and $\lambda, \mu \in \mathbb{R}$.*

The simplest examples of semi-Riemannian manifolds satisfying (I) are indefinite complex space forms and paracomplex space forms. (See [4], [42] and [54] for details and further references.)

It is well-known that, not only the eigenvalues, but also the existence of distinguished eigenspaces of Jacobi operators provide certain geometric information. In the following, we make an additional hypothesis on the eigenspaces of Jacobi operators. To describe this property, we introduce the following useful notation. For each $x \in S_p(M)$, let $E_\lambda(x)$ denote the direct sum of the subspace generated by x and eigenspace associated to the eigenvalue $\varepsilon_x \lambda$ of the Jacobi operator, that is $E_\lambda(x) = \text{span}\{x\} \oplus ker(R_x - \varepsilon_x \lambda id)$. Also let

(II) *For each $x \in S_p(M)$, if z is a unit vector in $E_\lambda(x)$ then $E_\lambda(x) = E_\lambda(z)$, and moreover, if $y \in ker(R_x - \varepsilon_x \mu id)$ then $x \in ker(R_y - \varepsilon_y \mu id)$.*

If (M, g, J) is a nonflat indefinite complex or paracomplex space form, the Jacobi operator with respect to each $x \in S(M)$ has exactly two distinct eigenvalues λ and $\lambda/4$. Moreover, $E_\lambda(x) = \text{span}\{x, Jx\}$ for each $x \in S(M)$, where J denotes the complex or paracomplex structure of M, and (II) follows trivially from the fact $g(x, Jy) + g(Jx, y) = 0$ for every $x, y \in TM$. Furthermore, note that $E_\lambda(x)$ has induced definite inner product for each $x \in S(M)$ if (M, g, J) is an indefinite complex space form. But it is a Lorentzian subspace of $T_p M$ in the paracomplex case, at each $p \in M$.

For the sake of briefness, we state the following definition.

Definition 5.2.1. *A semi-Riemannian manifold (M, g) is said to be special Osserman if* (I) *and* (II) *above are satisfied at each $p \in M$.*

Note that, as a consequence of Theorem 1.3.1, if (M, g) is a semi-Riemannian special Osserman manifold then the eigenvalues $\varepsilon_x \lambda$ and $\varepsilon_x \mu$ of Jacobi operators are constant on $S^\pm(M)$, provided that M is connected and $\dim M > 4$.

The purpose of this section is to prove the following result on the characterization of semi-Riemannian special Osserman manifolds.

Theorem 5.2.1. *Let (M, g) be a complete and simply connected semi-Riemannian special Osserman manifold. Then it is isometric to one of the following:*

(a) *an indefinite complex space form with metric tensor of signature $(2k, 2r)$, $k, r \geq 0$,*

(b) *an indefinite quaternionic space form with metric tensor of signature $(4k, 4r)$, $k, r \geq 0$,*

(c) *a paracomplex space form with metric tensor of signature (k, k),*

(d) *a paraquaternionic space form with metric tensor of signature $(2k, 2k)$, or*

(e) *a Cayley plane over the octaves with definite or indefinite metric tensor, or a Cayley plane over the anti-octaves with indefinite metric tensor of signature $(8, 8)$.*

Remark 5.2.1. Note that (I) and (II) reduces to (1), (2) in subsection 2.2.2 if the metric tensor g is assumed to be positive definite. Moreover, Theorem 5.2.1 reduces to Theorem 2.2.4 for Riemannian metric tensors.

We prove Theorem 5.2.1 in three steps. The first one consists of the explicit determination of the curvature tensor corresponding to a semi-Riemannian manifold that is special Osserman. This is worked out in Theorem 5.3.2 when the multiplicity of the distinguished eigenvalue λ is different from 7 or 15. In the second step, by making a repeated use of the second Bianchi identity, it is shown that such semi-Riemannian special Osserman manifolds are locally symmetric and, a local version of Theorem 5.2.1 is proved when the multiplicity of λ is different from 7 and 15. Finally, in the third step, we show the nonexistence of semi-Riemannian special Osserman manifolds with eigenvalue λ of multiplicity 15 and, when the multiplicity of λ equals 7, they correspond to the Cayley planes.

5.3 Classification of Semi-Riemannian Special Osserman Manifolds

In this section we restrict our attention to the tangent space of a semi-Riemannian special Osserman manifold (M, g) at $p \in M$. We use (I) and (II) to prove some properties and technical results which we need later.

It is important to note that the subspace $E_\mu(x)$ does not satisfy a condition similar to that imposed on $E_\lambda(x)$ in (II). This shows that it is not possible to interchange the distinct eigenvalues $\varepsilon_x \lambda$ and $\varepsilon_x \mu$ in the definition of a semi-Riemannian special Osserman manifold since their associated eigenspaces play different roles. Note also that it immediately follows from (II) that if $y \in ker(R_x - \varepsilon_x \lambda id)$, then $x \in ker(R_y - \varepsilon_y \lambda id)$. We begin with the following:

Lemma 5.3.1. *Let (M, g) be a semi-Riemannian special Osserman manifold. Then,*

(i) *If $x, y \in S_p(M)$, with $y \in E_\lambda(x)^\perp$ then $y \in ker(R_x - \mu \varepsilon_x id)$.*
(ii) *If x, y are chosen as in (i) then $E_\lambda(y) \perp E_\lambda(x)$.*
(iii) *$T_p M$ can be decomposed as a direct sum of the orthogonal subspaces $E_\lambda(\cdot)$.*

Proof. (i) follows immediately from the decomposition $T_p M = E_\lambda(x) \oplus ker(R_x - \mu \varepsilon_x id)$ since y is assumed to be orthogonal to $E_\lambda(x)$. To prove (ii), let $\bar{x}$ be a unit vector in $E_\lambda(x)$. Then (II) implies that $E_\lambda(\bar{x})$ coincides with $E_\lambda(x)$ and, it follows from (i) that $y \in ker(R_{\bar{x}} - \mu \varepsilon_{\bar{x}} id)$. Then, again by (II), we have $\bar{x} \in ker(R_y - \mu \varepsilon_y id)$, and thus $\bar{x} \perp E_\lambda(y)$ for all unit $\bar{x} \in E_\lambda(x)$. Finally, (iii) is obtained as a direct application of (ii). $\qquad \square$

Note that (iii) in the previous lemma gives an orthogonal decomposition of the tangent space as a direct sum of $E_\lambda(\cdot)$. Moreover, the restriction of the metric tensor to each one of these subspaces is nondegenerate, but its signature may change from one subspace to another. Later, we will show that there are only two possibilities for the signature of the induced inner product on $E_\lambda(\cdot)$. That is, the signature is either definite or neutral (cf. Proposition 5.3.1).

Lemma 5.3.2. *Considering the decomposition of T_pM,*

$$T_pM = E_\lambda(x) \oplus E_\lambda(y) \oplus E_\lambda(z) \oplus \cdots, \tag{5.8}$$

given by the previous lemma, the curvature tensor of (M, g) satisfies

(i) $R(y, x_1)x_2 = -R(y, x_2)x_1 = -\dfrac{1}{2}R(x_1, x_2)y,$

(ii) $R(x, y)z = 0,$

(iii) $R(x_1, x_2)x_3 = 0,$

where x_1, x_2, x_3 are orthogonal unit vectors in $E_\lambda(x)$.

Proof. Let x_1, x_2 be orthogonal unit vectors in $E_\lambda(x)$ and take a, b nonzero real numbers such that $w = ax_1 + bx_2$ is a unit vector. Then by (II), since $E_\lambda(w) = E_\lambda(x)$, y is orthogonal to $E_\lambda(w)$. Therefore $R_w(y) = \mu\varepsilon_w y$, and hence

$$\begin{aligned}
\mu\varepsilon_w y &= R(y, ax_1 + bx_2)(ax_1 + bx_2) \\
&= a^2 R(y, x_1)x_1 + b^2 R(y, x_2)x_2 + ab(R(y, x_1)x_2 + R(y, x_2)x_1) \\
&= a^2 \mu\varepsilon_{x_1} y + b^2 \mu\varepsilon_{x_2} y + ab(R(y, x_1)x_2 + R(y, x_2)x_1) \\
&= \mu(a^2\varepsilon_{x_1} + b^2\varepsilon_{x_2})y + ab(R(y, x_1)x_2 + R(y, x_2)x_1) \\
&= \mu\varepsilon_w y + ab(R(y, x_1)x_2 + R(y, x_2)x_1).
\end{aligned}$$

Also, since $ab \neq 0$, we have $R(y, x_1)x_2 = -R(y, x_2)x_1$. Now, by the first Bianchi identity,

$$\begin{aligned}
R(y, x_1)x_2 &= -R(x_1, x_2)y - R(x_2, y)x_1 = -R(x_1, x_2)y + R(y, x_2)x_1 \\
&= -R(x_1, x_2)y - R(y, x_1)x_2,
\end{aligned}$$

and hence $R(y, x_1)x_2 = -\dfrac{1}{2}R(x_1, x_2)y$, in proving *(i)*.

To prove *(ii)*, again take a, b nonzero real numbers such that $w = ay + bz$ is a unit vector. Then since $w \in E_\lambda(x)^\perp$, $R_w(x) = \mu\varepsilon_w x$ and hence

$$\begin{aligned}
\mu\varepsilon_w x &= R(x, ay + bz)(ay + bz) \\
&= a^2 R(x, y)y + b^2 R(x, z)z + ab(R(x, y)z + R(x, z)y) \\
&= a^2 \mu\varepsilon_y x + b^2 \mu\varepsilon_z x + ab(R(x, y)z + R(x, z)y) \\
&= \mu(a^2\varepsilon_y + b^2\varepsilon_z)x + ab(R(x, y)z + R(x, z)y) \\
&= \mu\varepsilon_w x + ab(R(x, y)z + R(x, z)y).
\end{aligned}$$

Thus $R(x,y)z = -R(x,z)y$. Similarly, it can be shown that $R(y,z)x = -R(y,x)z$ and $R(z,x)y = -R(z,y)x$. Now, from the first Bianchi identity,

$$R(x,y)z = -R(y,z)x - R(z,x)y = R(y,x)z + R(x,z)y$$
$$= -R(x,y)z - R(x,y)z = -2R(x,y)z,$$

and hence $R(x,y)z = 0$, proving (ii).

We prove (iii) with a similar argument. Let a, b be nonzero real numbers such that $w = ax_1 + bx_2$ is a unit vector. Then we have $R_w(x_3) = \lambda\varepsilon_w x_3$ since x_3 is orthogonal to w and belongs to $E_\lambda(w) = E_\lambda(x)$. Once more, by a linearization process,

$$\lambda\varepsilon_w x_3 = R(x_3, ax_1 + bx_2)(ax_1 + bx_2)$$
$$= a^2 R(x_3, x_1)x_1 + b^2 R(x_3, x_2)x_2 + ab(R(x_3, x_1)x_2 + R(x_3, x_2)x_1)$$
$$= a^2 \lambda\varepsilon_{x_1} x_3 + b^2 \lambda\varepsilon_{x_2} x_3 + ab(R(x_3, x_1)x_2 + R(x_3, x_2)x_1)$$
$$= \lambda(a^2\varepsilon_{x_1} + b^2\varepsilon_{x_2})x_3 + ab(R(x_3, x_1)x_2 + R(x_3, x_2)x_1)$$
$$= \lambda\varepsilon_w x_3 + ab(R(x_3, x_1)x_2 + R(x_3, x_2)x_1),$$

and thus $R(x_3, x_1)x_2 = -R(x_3, x_2)x_1$. Now $R(x_1, x_2)x_3 = -R(x_1, x_3)x_2$ and $R(x_2, x_3)x_1 = -R(x_2, x_1)x_3$ can be obtained similarly and it follows that

$$R(x_1, x_2)x_3 = -R(x_2, x_3)x_1 - R(x_3, x_1)x_2 = R(x_2, x_1)x_3 + R(x_1, x_3)x_2$$
$$= -R(x_1, x_2)x_3 - R(x_1, x_2)x_3 = -2R(x_1, x_2)x_3.$$

This shows that $R(x_1, x_2)x_3 = 0$, in completing the proof.   □

From now on, we denote by τ the multiplicity of the distinguished eigenvalue λ. Therefore $E_\lambda(\xi_0)$ is a $(\tau+1)$-dimensional subspace of T_pM, and we write $E_\lambda(\xi_0) = span\{\xi_0, \xi_1, \ldots, \xi_\tau\}$ for each $\xi_0 \in S_p(M)$, where $\{\xi_1, \ldots, \xi_\tau\}$ is an orthonormal basis for $ker(R_{\xi_0} - \lambda\varepsilon_{\xi_0} id)$.

Next we determine the Jacobi operator R_w for $w \in S_p(M)$ of the form $w = ax_0 + by_0$, where $y_0 \in E_\lambda(x_0)^\perp$. From the form of such a Jacobi operator, an important formula is derived in Lemma 5.3.3. The eigenspaces $ker(R_w - \lambda\varepsilon_w id)$ and $ker(R_w - \mu\varepsilon_w id)$ are also described in Lemma 5.3.4. These results are used extensively in the next sections.

Lemma 5.3.3. *Let x_0, $y_0 \in S_p(M)$ with $y_0 \in E_\lambda(x_0)^\perp$. The relation*

$$\sum_{j=1}^{\tau} R(x_i, y_0, x_0, y_j)R(x_k, y_0, x_0, y_j)\varepsilon_{y_j} = \delta_{ik}\left(\frac{\lambda - \mu}{3}\right)^2 \varepsilon_{x_0}\varepsilon_{x_i}\varepsilon_{y_0}$$

holds for all $i, k = 1, \ldots, \tau$, where $E_\lambda(x_0) = span\{x_0, x_1, \ldots, x_\tau\}$, $E_\lambda(y_0) = span\{y_0, y_1, \ldots, y_\tau\}$ and δ_{ik} denotes the Kronecker's delta.

Proof. Let x_0, $y_0 \in S_p(M)$ with $y_0 \in E_\lambda(x_0)^\perp$ and take nonzero a, b such that $w = ax_0 + by_0$ is a unit vector. Associated with the decomposition

$$T_pM = E_\lambda(x_0) \oplus E_\lambda(y_0) \oplus (E_\lambda(x_0) \oplus E_\lambda(y_0))^\perp,$$

consider an orthonormal basis $\{x_0, x_1, \ldots, x_\tau, y_0, y_1, \ldots, y_\tau, z_1, \ldots, z_{n-2(\tau+1)}\}$ for T_pM. First we analyze how the Jacobi operator R_w acts on each vector in this basis below.

$$R_w(x_0) = R(x_0, ax_0 + by_0)(ax_0 + by_0) = abR(x_0, y_0)x_0 + b^2 R(x_0, y_0)y_0$$
$$= -ab\mu\varepsilon_{x_0}y_0 + b^2\mu\varepsilon_{y_0}x_0.$$

$$R_w(x_i) = R(x_i, ax_0 + by_0)(ax_0 + by_0)$$
$$= a^2 R(x_i, x_0)x_0 + b^2 R(x_i, y_0)y_0 + ab(R(x_i, x_0)y_0 + R(x_i, y_0)x_0)$$
$$= a^2 \lambda\varepsilon_{x_0}x_i + b^2\mu\varepsilon_{y_0}x_i + ab(R(x_i, x_0)y_0 + R(x_i, y_0)x_0)$$
$$= (a^2\varepsilon_{x_0}\lambda + b^2\varepsilon_{y_0}\mu)x_i + ab(R(x_i, y_0)x_0 + R(x_i, x_0)y_0).$$

Now, since $R(x_i, x_0)y_0 = 2R(x_i, y_0)x_0$, by (i) of Lemma 5.3.2, the above expression reduces to $R_w(x_i) = (a^2\varepsilon_{x_0}\lambda + b^2\varepsilon_{y_0}\mu)x_i + 3abR(x_i, y_0)x_0$. Moreover, since $R(x_i, y_0)x_0$ is in $span\{y_1, \ldots, y_\tau\}$, we have

$$R(x_i, y_0)x_0 = \sum_{j=1}^{\tau} R(x_i, y_0, x_0, y_j)\varepsilon_{y_j}y_j$$

and thus,

$$R_w(x_i) = (a^2\varepsilon_{x_0}\lambda + b^2\varepsilon_{y_0}\mu)x_i + 3ab\sum_{j=1}^{\tau} R(x_i, y_0, x_0, y_j)\varepsilon_{y_j}y_j, \quad (i = 1, \ldots, \tau).$$

$$R_w(y_0) = R(y_0, ax_0 + by_0)(ax_0 + by_0) = a^2 R(y_0, x_0)x_0 + abR(y_0, x_0)y_0$$
$$= a^2\mu\varepsilon_{x_0}y_0 - ab\mu\varepsilon_{y_0}x_0.$$

$$R_w(y_i) = R(y_i, ax_0 + by_0)(ax_0 + by_0)$$
$$= a^2 R(y_i, x_0)x_0 + b^2 R(y_i, y_0)y_0 + ab(R(y_i, x_0)y_0 + R(y_i, y_0)x_0)$$
$$= a^2\mu\varepsilon_{x_0}y_i + b^2\lambda\varepsilon_{y_0}y_i + ab(R(y_i, x_0)y_0 + R(y_i, y_0)x_0)$$
$$= (a^2\varepsilon_{x_0}\mu + b^2\varepsilon_{y_0}\lambda)y_i + ab(R(y_i, x_0)y_0 + R(y_i, y_0)x_0).$$

Also, from Lemma 5.3.2-(i), $R_w(y_i) = (a^2\varepsilon_{x_0}\mu + b^2\varepsilon_{y_0}\lambda)y_i + 3abR(y_i, x_0)y_0$ and moreover, since $R(y_i, x_0)y_0 \in span\{x_1, \ldots, x_\tau\}$, we have $R(y_i, x_0)y_0 =$
$$\sum_{j=1}^{\tau} R(y_i, x_0, y_0, x_j)\varepsilon_{x_j}x_j = \sum_{j=1}^{\tau} R(x_j, y_0, x_0, y_i)\varepsilon_{x_j}x_j. \text{ Then}$$

$$R_w(y_i) = 3ab\sum_{j=1}^{\tau} R(x_j, y_0, x_0, y_i)\varepsilon_{x_j} x_j + (a^2\varepsilon_{x_0}\mu + b^2\varepsilon_{y_0}\lambda)y_i, \quad (i = 1, \ldots, \tau).$$

Now, take $z_i \in (E_\lambda(x_0) \oplus E_\lambda(y_0))^\perp$. Then

$$
\begin{aligned}
R_w(z_i) &= R(z_i, ax_0 + by_0)(ax_0 + by_0) \\
&= a^2 R(z_i, x_0)x_0 + b^2 R(z_i, y_0)y_0 + ab(R(z_i, x_0)y_0 + R(z_i, y_0)x_0) \\
&= a^2\mu\varepsilon_{x_0} z_i + b^2\mu\varepsilon_{y_0} z_i + ab(R(z_i, x_0)y_0 + R(z_i, y_0)x_0) \\
&= \mu(a^2\varepsilon_{x_0} + b^2\varepsilon_{y_0})z_i + ab(R(z_i, x_0)y_0 + R(z_i, y_0)x_0) \\
&= \mu\varepsilon_w z_i + ab(R(z_i, x_0)y_0 + R(z_i, y_0)x_0),
\end{aligned}
$$

since $R(z_i, x_0)y_0 = R(z_i, y_0)x_0 = 0$, $R_w(z_i) = \mu\varepsilon_w z_i$.

Now we consider the eigenspace of R_w associated to the eigenvalue $\lambda\varepsilon_w$. Since $\{b\varepsilon_{x_0}\varepsilon_{y_0} x_0 - ay_0, x_1, \ldots, x_\tau, y_1, \ldots, y_\tau, z_1, \ldots, z_{n-2(\tau+1)}\}$ is an orthonormal basis for $(span\{w\})^\perp$, for each $\xi \in ker(R_w - \lambda\varepsilon_w id)$, we can write

$$\xi = \alpha(b\varepsilon_{x_0}\varepsilon_{y_0} x_0 - ay_0) + \sum_{i=1}^{\tau}(\gamma_i x_i + \delta_i y_i) + \sum_{j=1}^{n-2(\tau+1)} \beta_j z_j \qquad (5.9)$$

and, by the previous computations, we obtain

$$R_w(\xi) = \alpha\mu\varepsilon_w(b\varepsilon_{x_0}\varepsilon_{y_0} x_0 - ay_0) + \sum_{i=1}^{\tau}(\gamma_i' x_i + \delta_i' y_i) + \sum_{j=1}^{n-2(\tau+1)} \beta_j\mu\varepsilon_w z_j.$$

If ξ is assumed to be an eigenvector of R_w associated to λ, (5.9) shows that

$$R_w(\xi) = \alpha\lambda\varepsilon_w(b\varepsilon_{x_0}\varepsilon_{y_0} x_0 - ay_0) + \sum_{i=1}^{\tau}(\gamma_i\lambda\varepsilon_w x_i + \delta_j\lambda\varepsilon_w y_i) + \sum_{j=1}^{n-2(\tau+1)} \beta_j\lambda\varepsilon_w z_j,$$

and thus

$$\alpha(\lambda - \mu)\varepsilon_w(b\varepsilon_{x_0}\varepsilon_{y_0} x_0 - ay_0) + \sum_{i=1}^{\tau}(\gamma_i'' x_i + \delta_i'' y_i) + \sum_{j=1}^{n-2(\tau+1)} \beta_j(\lambda - \mu)\varepsilon_w z_j = 0.$$

Also, since $\{b\varepsilon_{x_0}\varepsilon_{y_0} x_0 - ay_0, x_1, \ldots, x_\tau, y_1, \ldots, y_\tau, z_1, \ldots, z_{n-2(\tau+1)}\}$ is a basis for $T_p M$, $\alpha = 0$ and $\beta_j = 0$, $j = 1, \ldots, n - 2(\tau + 1)$, and therefore (5.9) implies that $\xi \in span\{x_1, \ldots, x_\tau, y_1, \ldots, y_\tau\}$. Hence, if V denotes the subspace of $T_p M$ spanned by $\{x_1, \ldots, x_\tau, y_1, \ldots, y_\tau\}$, we have

$$ker(R_w - \lambda\varepsilon_w id) \subset V. \qquad (5.10)$$

Now, let us denote by $\tilde{R}_w$ the restriction of R_w to V. Then, the previously obtained expression of R_w shows that the matrix of $\tilde{R}_w$ with respect to the basis $\{x_1, \ldots, x_\tau, y_1, \ldots, y_\tau\}$ is

$$\tilde{R}_w = \left(\begin{array}{c|c} (a^2\varepsilon_{x_0}\lambda + b^2\varepsilon_{y_0}\mu)Id_\tau & 3abC \\ \hline \\ 3abB & (a^2\varepsilon_{x_0}\mu + b^2\varepsilon_{y_0}\lambda)Id_\tau \end{array} \right), \tag{5.11}$$

where B and C are the $(\tau \times \tau)$-matrices given by $(B_{ij}) = (R(x_j, y_0, x_0, y_i)\varepsilon_{y_i})$ and $(C_{ij}) = (R(x_i, y_0, x_0, y_j)\varepsilon_{x_i})$, respectively.

Note at this point that, since $ker(R_w - \lambda\varepsilon_w id) \subset V$, $\tilde{R}_w$ has two eigenvalues, λ and μ, both with the same multiplicity τ. Now, from (5.11),

$$\tilde{R}_w - \lambda\varepsilon_w Id_{2\tau} = \left(\begin{array}{c|c} -b^2(\lambda - \mu)\varepsilon_{y_0}Id_\tau & 3abC \\ \hline \\ 3abB & -a^2(\lambda - \mu)\varepsilon_{x_0}Id_\tau \end{array} \right),$$

and thus, $ker(\tilde{R}_w - \lambda\varepsilon_w id)$ is determined by

$$\begin{cases} -b^2(\lambda - \mu)\varepsilon_{y_0}\mathbf{x} + 3abC\mathbf{y} = 0 \\ 3abB\mathbf{x} - a^2(\lambda - \mu)\varepsilon_{x_0}\mathbf{y} = 0 \end{cases} \tag{5.12}$$

where $\mathbf{x}, \mathbf{y}$ are vectors in $span\{x_1, \ldots, x_\tau\}$ and $span\{y_1, \ldots, y_\tau\}$, respectively. One can directly check that the solution of (5.12) is given by

$$\mathbf{y} = \frac{3}{\lambda - \mu}\frac{b}{a}\varepsilon_{x_0}B\mathbf{x} \quad \text{and} \quad CB\mathbf{x} = \left(\frac{\lambda - \mu}{3}\right)^2 \varepsilon_{x_0}\varepsilon_{y_0}\mathbf{x}.$$

Also, since $ker(\tilde{R}_w - \lambda\varepsilon_w id)$ has dimension τ and $\mathbf{x}, \mathbf{y}$ are $(\tau \times 1)$-matrices, it follows that $CB\mathbf{x} = \left(\frac{\lambda - \mu}{3}\right)^2 \varepsilon_{x_0}\varepsilon_{y_0}\mathbf{x}$ holds for all vectors $\mathbf{x}$ in $span\{x_1, \ldots, x_\tau\}$. Therefore,

$$CB = \left(\frac{\lambda - \mu}{3}\right)^2 \varepsilon_{x_0}\varepsilon_{y_0}Id_\tau. \tag{5.13}$$

Finally note that the vector in the position (i, k) of the product CB is given by $\varepsilon_{x_i}\sum_{j=1}^{\tau} R(x_i, y_0, x_0, y_j)R(x_k, y_0, x_0, y_j)\varepsilon_{y_j}$, and from (5.13), we obtain

$$\sum_{j=1}^{\tau} R(x_i, y_0, x_0, y_j)R(x_k, y_0, x_0, y_j)\varepsilon_{y_j} = \delta_{ik}\left(\frac{\lambda - \mu}{3}\right)^2 \varepsilon_{x_0}\varepsilon_{x_i}\varepsilon_{y_0}$$

for $i, k = 1, \ldots, \tau$, which completes the proof. $\square$

Now we study the eigenspaces corresponding to the eigenvalues $\lambda \varepsilon_w$ and $\mu \varepsilon_w$ of the Jacobi operator associated to $w \in S_p(M)$ as in the previous lemma.

Lemma 5.3.4. *Let $x_0, y_0 \in S_p(M)$, with $y_0 \in E_\lambda(x_0)^\perp$, and a, b are nonzero real numbers such that $w = ax_0 + by_0$ is a unit vector. Then,*

$$u_i = x_i - \frac{3}{2(\lambda - \mu)} \frac{b}{a} \varepsilon_{x_0} R(x_0, x_i) y_0, \qquad i = 1, \ldots, \tau,$$

$$v_i = \frac{b}{a} \varepsilon_{y_0} x_i + \frac{3}{2(\lambda - \mu)} R(x_0, x_i) y_0, \qquad i = 1, \ldots, \tau,$$

satisfy

(i) $\{u_1, \ldots, u_\tau\}$ is a basis for $ker(R_w - \lambda \varepsilon_w id)$,
(ii) $span\{v_1, \ldots, v_\tau\}$ is a τ-dimensional subspace of $ker(R_w - \mu \varepsilon_w id)$,
(iii) $R_{u_i} v_j = \mu g(u_i, u_i) v_j$, for $i, j = 1, \ldots, \tau$,

where $\{x_1, \ldots, x_\tau\}$ is an orthonormal basis for $ker(R_{x_0}^. - \lambda \varepsilon_{x_0} id)$.

Proof. Let $x_0, y_0 \in S_p(M)$ with $y_0 \in E_\lambda(x_0)^\perp$ and take nonzero a, b such that $w = ax_0 + by_0$ is a unit vector. If $\{y_1, \ldots, y_\tau\}$ is an orthonormal basis for $ker(R_{y_0} - \lambda \varepsilon_{y_0} id)$, we have by the previous lemma that

$$R_w(x_i) = (a^2 \varepsilon_{x_0} \lambda + b^2 \varepsilon_{y_0} \mu) x_i - \frac{3}{2} ab R(x_0, x_i) y_0 \qquad (i = 1, \ldots, \tau) \quad (5.14)$$

and

$$R_w(y_i) = (a^2 \varepsilon_{x_0} \mu + b^2 \varepsilon_{y_0} \lambda) y_i + 3ab R(y_i, x_0) y_0 \qquad (i = 1, \ldots, \tau). \quad (5.15)$$

Now we study the operator R_w acting on $R(x_0, x_i) y_0$, $i = 1, \ldots, \tau$. Since

$$R(x_0, x_r) y_0 = \sum_{j=1}^{\tau} R(x_0, x_r, y_0, y_j) \varepsilon_{y_j} y_j, \qquad r = 1, \ldots, \tau, \quad (5.16)$$

it follows that $R_w(R(x_0, x_i) y_0) = \sum_{j=1}^{\tau} R(x_0, x_i, y_0, y_j) \varepsilon_{y_j} R_w(y_j)$, and from (5.15),

$$R_w(R(x_0, x_i) y_0) = (a^2 \varepsilon_{x_0} \mu + b^2 \varepsilon_{y_0} \lambda) \sum_{j=1}^{\tau} R(x_0, x_i, y_0, y_j) \varepsilon_{y_j} y_j$$

$$+ 3ab \sum_{j=1}^{\tau} R(x_0, x_i, y_0, y_j) \varepsilon_{y_j} R(y_j, x_0) y_0.$$

Then, (5.16) and $R(y_j, x_0)y_0 = \sum_{k=1}^{\tau} R(y_j, x_0, y_0, x_k)\varepsilon_{x_k} x_k$ imply that

$$R_w(R(x_0, x_i)y_0) = (a^2\varepsilon_{x_0}\mu + b^2\varepsilon_{y_0}\lambda)R(x_0, x_i)y_0$$

$$+3ab\sum_{k=1}^{\tau}\left\{\sum_{j=1}^{\tau} R(x_0, x_i, y_0, y_j)R(y_j, x_0, y_0, x_k)\varepsilon_{y_j}\right\}\varepsilon_{x_k} x_k.$$

Using $R(x_0, x_i, y_0, y_j) = -2R(x_i, y_0, x_0, y_j)$ and $R(y_j, x_0, y_0, x_k) = R(x_k, y_0, x_0, y_j)$, the previous expression reduces to

$$R_w(R(x_0, x_i)y_0) = (a^2\varepsilon_{x_0}\mu + b^2\varepsilon_{y_0}\lambda)R(x_0, x_i)y_0$$

$$-6ab\sum_{k=1}^{\tau}\left\{\sum_{j=1}^{\tau} R(x_i, y_0, x_0, y_j)R(x_k, y_0, x_0, y_j)\varepsilon_{y_j}\right\}\varepsilon_{x_k} x_k.$$

Hence, from Lemma 5.3.3, it follows that

$$R_w(R(x_0, x_i)y_0) = (a^2\varepsilon_{x_0}\mu + b^2\varepsilon_{y_0}\lambda)R(x_0, x_i)y_0$$

$$-6ab\sum_{k=1}^{\tau}\delta_{ik}\left(\frac{\lambda-\mu}{3}\right)^2\varepsilon_{x_0}\varepsilon_{x_i}\varepsilon_{y_0}\varepsilon_{x_k} x_k,$$

where

$$R_w(R(x_0, x_i)y_0) = (a^2\varepsilon_{x_0}\mu + b^2\varepsilon_{y_0}\lambda)R(x_0, x_i)y_0 - \frac{2}{3}ab(\lambda-\mu)^2\varepsilon_{x_0}\varepsilon_{y_0}x_i.$$

$$(5.17)$$

Now, from (5.14) and (5.17),

$$R_w(u_i) = R_w(x_i) - \frac{3}{2(\lambda-\mu)}\frac{b}{a}\varepsilon_{x_0}R_w(R(x_0, x_i)y_0)$$

$$= (a^2\varepsilon_{x_0}\lambda + b^2\varepsilon_{y_0}\mu)x_i - a\frac{3}{2}bR(x_0, x_i)y_0$$

$$-(a^2\varepsilon_{x_0}\mu + b^2\varepsilon_{y_0}\lambda)\frac{3}{2(\lambda-\mu)}\frac{b}{a}\varepsilon_{x_0}R(x_0, x_i)y_0 + (b^2\varepsilon_{y_0}\lambda - b^2\varepsilon_{y_0}\mu)x_i$$

$$= (a^2\varepsilon_{x_0}\lambda + b^2\varepsilon_{y_0}\mu)x_i - (a^2\varepsilon_{x_0}\lambda - a^2\varepsilon_{x_0}\mu)\frac{3}{2(\lambda-\mu)}\frac{b}{a}\varepsilon_{x_0}R(x_0, x_i)y_0$$

$$-(a^2\varepsilon_{x_0}\mu + b^2\varepsilon_{y_0}\lambda)\frac{3}{2(\lambda-\mu)}\frac{b}{a}\varepsilon_{x_0}R(x_0, x_i)y_0 + (b^2\varepsilon_{y_0}\lambda - b^2\varepsilon_{y_0}\mu)x_i$$

$$= \lambda(a^2\varepsilon_{x_0} + b^2\varepsilon_{y_0})x_i - \lambda(a^2\varepsilon_{x_0} + b^2\varepsilon_{y_0})\frac{3}{2(\lambda-\mu)}\frac{b}{a}\varepsilon_{x_0}R(x_0, x_i)y_0$$

$$= \lambda\varepsilon_w u_i, \quad (i = 1, \dots, \tau).$$

$$R_w(v_i) = \frac{b}{a}\varepsilon_{y_0}R_w(x_i) + \frac{3}{2(\lambda - \mu)}R_w(R(x_0, x_i)y_0)$$

$$= (a^2\varepsilon_{x_0}\lambda + b^2\varepsilon_{y_0}\mu)\frac{b}{a}\varepsilon_{y_0}x_i - (b^2\varepsilon_{y_0})\frac{3}{2}R(x_0, x_i)y_0$$

$$+(a^2\varepsilon_{x_0}\mu + b^2\varepsilon_{y_0}\lambda)\frac{3}{2(\lambda - \mu)}R(x_0, x_i)y_0 - (a\varepsilon_{x_0}\lambda - a\varepsilon_{x_0}\mu)b\varepsilon_{y_0}x_i$$

$$= (a^2\varepsilon_{x_0}\lambda + b^2\varepsilon_{y_0}\mu)\frac{b}{a}\varepsilon_{y_0}x_i - (b^2\varepsilon_{y_0}\lambda - b^2\varepsilon_{y_0}\mu)\frac{3}{2(\lambda - \mu)}R(x_0, x_i)y_0$$

$$+(a^2\varepsilon_{x_0}\mu + b^2\varepsilon_{y_0}\lambda)\frac{3}{2(\lambda - \mu)}R(x_0, x_i)y_0 - (a^2\varepsilon_{x_0}\lambda - a^2\varepsilon_{x_0}\mu)\frac{b}{a}\varepsilon_{y_0}x_i$$

$$= \mu(a^2\varepsilon_{x_0} + b^2\varepsilon_{y_0})\frac{b}{a}\varepsilon_{y_0}x_i + \mu(a^2\varepsilon_{x_0} + b^2\varepsilon_{y_0})\frac{3}{2(\lambda - \mu)}R(x_0, x_i)y_0$$

$$= \mu\varepsilon_w v_i, \quad (i = 1, \ldots, \tau).$$

The expressions above show that $span\{u_1, \ldots, u_\tau\}$ and $span\{v_1, \ldots, v_\tau\}$ are subspaces of $ker(R_w - \lambda\varepsilon_w id)$ and $ker(R_w - \mu\varepsilon_w id)$, respectively. To complete the proof, we need to show that both are sets of orthogonal nonnull vectors. For this, note that (5.16) and Lemma 5.3.3 give

$$g(R(x_0, x_i)y_0, R(x_0, x_j)y_0)$$

$$= g\left(\sum_{k=1}^{\tau}R(x_0, x_i, y_0, y_k)\varepsilon_{y_k}y_k, \sum_{l=1}^{\tau}R(x_0, x_j, y_0, y_l)\varepsilon_{y_l}y_l\right)$$

$$= \sum_{k,l=1}^{\tau}(-2R(x_i, y_0, x_0, y_k))(-2R(x_j, y_0, x_0, y_l))\varepsilon_{y_k}\varepsilon_{y_l}g(y_k, y_l)$$

$$= 4\sum_{k,l=1}^{\tau}R(x_i, y_0, x_0, y_k)R(x_j, y_0, x_0, y_l)\varepsilon_{y_k}\varepsilon_{y_l}\delta_{kl}\varepsilon_{y_k}$$

$$= 4\sum_{k=1}^{\tau}R(x_i, y_0, x_0, y_k)R(x_j, y_0, x_0, y_k)\varepsilon_{y_k}$$

$$= 4\delta_{ij}\left(\frac{\lambda - \mu}{3}\right)^2\varepsilon_{x_0}\varepsilon_{x_i}\varepsilon_{y_0}, \quad (i, j = 1, \ldots, \tau).$$

Now, using this expression and the fact that $R(x_0, x_i)y_0$ and $R(x_0, x_j)y_0$ are orthogonal to the subspace $span\{x_1, \ldots, x_\tau\}$, one obtains

$g(u_i, u_j)$

$$= g\left(x_i - \frac{3}{2(\lambda - \mu)}\frac{b}{a}\varepsilon_{x_0}R(x_0, x_i)y_0, x_j - \frac{3}{2(\lambda - \mu)}\frac{b}{a}\varepsilon_{x_0}R(x_0, x_j)y_0\right)$$

$$= g(x_i, x_j) + \left(\frac{3}{2(\lambda - \mu)}\frac{b}{a}\right)^2 g(R(x_0, x_i)y_0, R(x_0, x_j)y_0)$$

$$= \delta_{ij}\varepsilon_{x_i} + \delta_{ij}\frac{b^2}{a^2}\varepsilon_{x_0}\varepsilon_{x_i}\varepsilon_{y_0}$$

$$= \delta_{ij}\frac{\varepsilon_{x_0}\varepsilon_{x_i}}{a^2}(a^2\varepsilon_{x_0} + b^2\varepsilon_{y_0})$$

$$= \delta_{ij}\frac{\varepsilon_{x_0}\varepsilon_{x_i}\varepsilon_w}{a^2}, \quad (i, j = 1, \ldots, \tau),$$

and

$g(v_i, v_j)$

$$= g\left(\frac{b}{a}\varepsilon_{y_0}x_i + \frac{3}{2(\lambda - \mu)}R(x_0, x_i)y_0, \frac{b}{a}\varepsilon_{y_0}x_j + \frac{3}{2(\lambda - \mu)}R(x_0, x_j)y_0\right)$$

$$= \frac{b^2}{a^2}g(x_i, x_j) + \left(\frac{3}{2(\lambda - \mu)}\right)^2 g(R(x_0, x_i)y_0, R(x_0, x_j)y_0)$$

$$= \frac{b^2}{a^2}\delta_{ij}\varepsilon_{x_i} + \delta_{ij}\varepsilon_{x_0}\varepsilon_{x_i}\varepsilon_{y_0}$$

$$= \delta_{ij}\frac{\varepsilon_{y_0}\varepsilon_{x_i}}{a^2}(a^2\varepsilon_{x_0} + b^2\varepsilon_{y_0})$$

$$= \delta_{ij}\frac{\varepsilon_{y_0}\varepsilon_{x_i}\varepsilon_w}{a^2}, \quad (i, j = 1, \ldots, \tau),$$

proving (i) and (ii).

To prove (iii), we choose $i, j \in \{1, \ldots, \tau\}$. Since u_i is nonnull, $\bar{u}_i = \frac{u_i}{\|u_i\|}$ belongs to $E_\lambda(w)$ by (i), and hence $E_\lambda(w) = E_\lambda(\bar{u}_i)$. On the other hand, (ii) implies that the vector v_j is orthogonal to the subspace $E_\lambda(w) = E_\lambda(\bar{u}_i)$, and hence, $v_j \in ker(R_{\bar{u}_i} - \mu\varepsilon_{\bar{u}_i}id)$. Thus, $R_{\bar{u}_i}v_j = \mu\varepsilon_{\bar{u}_i}v_j$ and the claim follows. $\square$

5.3.1 Multiplicities of the Eigenvalues

In this subsection, we study the possible multiplicities of the eigenvalue λ. To do this, we construct some linear maps on each subspace $E_\lambda(\cdot)$. The study of these maps provides information about the signature of the restriction of the metric tensor to the subspaces $E_\lambda(\cdot)$ and, as a consequence, we obtain the possible dimensions and signatures of the semi-Riemannian special Osserman metric tensors. Moreover, these tensors furnish the subspaces $E_\lambda(\cdot)$ with a certain Clifford module structure and, by using some topological restrictions on the existence of such structures, we obtain that the eigenvalue λ may only have the multiplicities of 1, 3, 7 or 15 (cf. Theorem 5.3.1.) We begin with the following definition.

Definition 5.3.1. *Let (M, g) be a semi-Riemannian special Osserman manifold. Let $x_0 \in S_p(M)$ and extend it to an orthonormal basis $\{x_0, x_1, \ldots, x_\tau\}$ for $E_\lambda(x_0)$. Then for each $i = 1, \ldots, \tau$, define $\phi_i : E_\lambda(x_0)^\perp \longrightarrow E_\lambda(x_0)^\perp$ by*

$$\phi_i \xi = \frac{3}{2(\lambda - \mu)} R(x_0, x_i)\xi, \qquad (5.18)$$

where $\xi \in E_\lambda(x_0)^\perp$.

Remark 5.3.1. Note that the maps ϕ_i are well-defined by Lemma 5.3.2-(*iii*). Moreover, if ξ_0 is a unit vector in $E_\lambda(x_0)^\perp$ with $E_\lambda(\xi_0) = span\{\xi_0, \xi_1, \ldots, \xi_\tau\}$, then since $R(x_0, x_i)\xi_0 \in span\{\xi_1, \ldots, \xi_\tau\}$, it follows that

$$\phi_i \xi_0 = \frac{3}{2(\lambda - \mu)} \sum_{j=1}^{\tau} R(x_0, x_i, \xi_0, \xi_j)\varepsilon_{\xi_j} \xi_j.$$

Now, from Lemma 5.3.2-(*i*), $R(x_0, x_i, \xi_0, \xi_j) = -2R(x_i, \xi_0, x_0, \xi_j)$ and hence

$$\phi_i \xi_0 = -\frac{3}{\lambda - \mu} \sum_{j=1}^{\tau} R(x_i, \xi_0, x_0, \xi_j)\varepsilon_{\xi_j} \xi_j. \qquad (5.19)$$

This shows that each subspace $E_\lambda(\cdot) \subset E_\lambda(x_0)^\perp$ remains invariant under the action of the ϕ_i's.

Now, by using Lemma 5.3.3, we have the following:

Lemma 5.3.5. *The linear maps defined in (5.3.1) satisfy*

(i) $g(\phi_i \xi, \eta) + g(\xi, \phi_i \eta) = 0$, for every $\xi, \eta \in E_\lambda(x_0)^\perp$,

(ii) $g(\phi_i \xi, \phi_j \xi) = \delta_{ij} \varepsilon_{x_0} \varepsilon_{x_i} g(\xi, \xi)$, for every $\xi \in E_\lambda(x_0)^\perp$,

(iii) $\phi_i \phi_j + \phi_j \phi_i = -2\delta_{ij} \varepsilon_{x_0} \varepsilon_{x_i} Id$,

where $i, j \in \{1, \ldots, \tau\}$.

Proof. If $\xi, \eta \in E_\lambda(x_0)^\perp$ then

$$g(\phi_i \xi, \eta) = \frac{3}{2(\lambda - \mu)} g(R(x_0, x_i)\xi, \eta) = \frac{3}{2(\lambda - \mu)} R(x_0, x_i, \xi, \eta)$$

$$= -\frac{3}{2(\lambda - \mu)} R(x_0, x_i, \eta, \xi) = -\frac{3}{2(\lambda - \mu)} g(R(x_0, x_i)\eta, \xi)$$

$$= -g(\phi_i \eta, \xi) = -g(\xi, \phi_i \eta),$$

proving (i). To prove (ii), let ξ_0 be a unit vector in $E_\lambda(x_0)^\perp$ and put $E_\lambda(\xi_0)$ $= span\{\xi_0, \xi_1, \ldots, \xi_\tau\}$. By (5.19), we have

$$g(\phi_i\xi_0, \phi_j\xi_0) = \left(\frac{3}{\lambda - \mu}\right)^2 \sum_{k=1}^\tau R(x_i, \xi_0, x_0, \xi_k)R(x_j, \xi_0, x_0, \xi_k)\varepsilon_{\xi_k},$$

and hence, Lemma 5.3.3 shows that $g(\phi_i\xi_0, \phi_j\xi_0) = \delta_{ij}\varepsilon_{x_0}\varepsilon_{x_i}g(\xi_0, \xi_0)$ for every unit vector ξ_0, proving (ii).

Finally we prove (iii). Let ξ be a vector in $E_\lambda(x_0)^\perp$ and note that, from (i) and (ii),

$$g(\phi_i\phi_j\xi, \xi) = -g(\phi_i\xi, \phi_j\xi) = -\delta_{ij}\varepsilon_{x_0}\varepsilon_{x_i}g(\xi, \xi). \qquad (5.20)$$

Now, if $\xi, \eta \in E_\lambda(x_0)^\perp$ then so is $(\xi + \eta)$, and (5.20) gives that $g(\phi_i\phi_j(\xi + \eta), \xi + \eta) = -\delta_{ij}\varepsilon_{x_0}\varepsilon_{x_i}g(\xi + \eta, \xi + \eta)$. After linearization

$$g(\phi_i\phi_j\xi, \xi) + g(\phi_i\phi_j\eta, \eta) + g(\phi_i\phi_j\xi, \eta) + g(\phi_i\phi_j\eta, \xi)$$
$$= -\delta_{ij}\varepsilon_{x_0}\varepsilon_{x_i}g(\xi, \xi) - \delta_{ij}\varepsilon_{x_0}\varepsilon_{x_i}g(\eta, \eta) - 2\delta_{ij}\varepsilon_{x_0}\varepsilon_{x_i}g(\xi, \eta),$$

and from (5.20), we obtain $g(\phi_i\phi_j\xi, \eta) + g(\phi_i\phi_j\eta, \xi) = -2\delta_{ij}\varepsilon_{x_0}\varepsilon_{x_i}g(\xi, \eta)$. Hence, the claimed result follows from (i). $\qquad \square$

Remark 5.3.2. Complex and paracomplex structures may only exist on even dimensional manifolds. This shows that the multiplicity of λ is necessarily odd. Moreover note that Hermitian metric tensors are of signature $(2k, 2r)$, $(k, r \geq 0)$. On the other hand para-Hermitian metric tensors are necessarily of neutral signature (k, k). As an application of this fact, we obtain the following.

Proposition 5.3.1. *Let $T_pM = E_\lambda(x) \oplus E_\lambda(y) \oplus \cdots$ be an orthogonal decomposition of the tangent space of a semi-Riemannian special Osserman manifold given in Lemma 5.3.1. Then for each $E_\lambda(\cdot)$, we have:*

(i) the restriction of the metric tensor to $E_\lambda(\cdot)$ is of signature $(\tau + 1, 0)$ or $(0, \tau + 1)$, or
(ii) the restriction of the metric tensor to $E_\lambda(\cdot)$ is of signature $\left(\frac{\tau+1}{2}, \frac{\tau+1}{2}\right)$.

Proof. Let $E_\lambda(\xi)$ be one of the subspaces in the decomposition (5.8) and take a unit $x_0 \in E_\lambda(\xi)^\perp$. Then, the linear maps $\phi_1, \ldots, \phi_\tau$ defined on $span\{x_0, x_1, \ldots, x_\tau\} = E_\lambda(x_0)$ leave $E_\lambda(\xi)$ invariant and Lemma 5.3.5 shows that they are complex or paracomplex structures. Further note that, if the restriction of the metric tensor is para-Hermitian for some ϕ_i's, then its signature is necessarily $\left(\frac{\tau+1}{2}, \frac{\tau+1}{2}\right)$. On the other hand, if the restriction of the metric tensor to $E_\lambda(\xi)$ is Hermitian with respect to all ϕ_i's, one can construct an orthonormal basis for $E_\lambda(\xi)$ by $\{\xi, \phi_1\xi, \ldots, \phi_\tau\xi\}$, and thus, the restriction of the metric tensor to $E_\lambda(\xi)$ is definite. $\qquad \square$

Remark 5.3.3. Note that, as a consequence of the proof of the previous Proposition, the restrictions of the metric tensor to the subspaces of the form $E_\lambda(\cdot)$ are of signatures $(\tau + 1, 0)$, $(0, \tau + 1)$ or $(\frac{\tau+1}{2}, \frac{\tau+1}{2})$. Indeed, if $\{x_0, \ldots, x_\tau\}$ is an orthonormal basis for $E_\lambda(x_0)$, then the induced ϕ_i's satisfy $\phi_i^2 = \sigma_i Id$, where $\sigma_i = -\varepsilon_{x_0}\varepsilon_{x_i}$, $i = 1, \ldots, \tau$. Therefore they are τ complex structures on $E_\lambda(x_0)^\perp$ or, exactly $(\frac{\tau-1}{2})$-complex and $(\frac{\tau+1}{2})$-paracomplex structures on $E_\lambda(x_0)^\perp$. Now, Proposition 5.3.1-(ii) shows that the first case in the above happens if the restriction of the metric tensor to $E_\lambda(x_0)$ is definite (and thus, this is the case for all $E_\lambda(\cdot)$) and the second case corresponds to the restricted metric tensor to $E_\lambda(x_0)$ of signature $(\frac{\tau+1}{2}, \frac{\tau+1}{2})$ (and thus this is the case for all $E_\lambda(\cdot)$.)

Next, we show that the multiplicity of λ strongly influences the dimension of a semi-Riemannian special Osserman manifold when it is assumed to be greater than 3.

Proposition 5.3.2. *Let (M, g) be a semi-Riemannian special Osserman manifold. If the multiplicity of λ is $\tau > 3$ then $dim M = 2(\tau + 1)$.*

Proof. Let $x_0, y_0 \in S_p(M)$ with $y_0 \in E_\lambda(x_0)^\perp$ and consider the linear maps $\phi_i : E_\lambda(x_0)^\perp \longrightarrow E_\lambda(x_0)^\perp$, $i = 1, \ldots, \tau$, defined by (5.3.1). It follows from the previous proposition that $\{\xi, \phi_1\xi, \ldots, \phi_\tau\xi\}$ is an orthonormal basis for $E_\lambda(y_0)$ for each unit $\xi \in E_\lambda(y_0)$, and thus we can write

$$\phi_i\phi_j\xi = \alpha_{ij}^0(\xi)\xi + \sum_{s=1}^{\tau}\alpha_{ij}^s(\xi)\phi_s\xi,$$

where $i \neq j$, $i, j \in \{1, \ldots, \tau\}$. Since $i \neq j$, $\phi_i\phi_j\xi \in (span\{\xi, \phi_i\xi, \phi_j\xi\})^\perp$, and the above expression reduces to

$$\phi_i\phi_j\xi = \sum_{s=1,\, s\neq i,j}^{\tau} \alpha_{ij}^s(\xi)\phi_s\xi. \tag{5.21}$$

Next, suppose that $dim M > 2(\tau + 1)$ and take a unit vector η orthogonal to the subspaces $E_\lambda(x_0)$ and $E_\lambda(y_0)$. Choose nonzero a, b in such a way that $w = a\xi + b\eta$ and $t = b\xi - a\varepsilon_\xi\varepsilon_\eta\eta$ are unit vectors. Then, for $l, m, n \in \{1, \ldots, \tau\}$ we have $g(\phi_l\phi_m\phi_n w, t) = 0$ since $\phi_l\phi_m\phi_n w \in E_\lambda(w)$ and $t \in E_\lambda(w)^\perp$. Hence,

$$0 = g(\phi_l\phi_m\phi_n(a\xi + b\eta), b\xi - a\varepsilon_\xi\varepsilon_\eta\eta)$$
$$= ab(g(\phi_l\phi_m\phi_n\xi, \xi) - \varepsilon_\xi\varepsilon_\eta g(\phi_l\phi_m\phi_n\eta, \eta))$$
$$-a^2\varepsilon_\xi\varepsilon_\eta g(\phi_l\phi_m\phi_n\xi, \eta) + b^2 g(\phi_l\phi_m\phi_n\eta, \xi).$$

Since $g(\phi_l\phi_m\phi_n\xi, \eta) = g(\phi_l\phi_m\phi_n\eta, \xi) = 0$ (note that $E_\lambda(\xi) \perp E_\lambda(\eta)$ and the both subspaces remain invariant under the action of the ϕ_i's), we obtain that

$g(\phi_l\phi_m\phi_n\xi,\xi) = \varepsilon_\xi\varepsilon_\eta g(\phi_l\phi_m\phi_n\eta,\eta)$. This shows that the coefficients $\alpha_{ij}^s(\xi)$ in (5.21) are given by

$$\begin{aligned}
\alpha_{ij}^s(\xi) &= g(\phi_i\phi_j\xi, \phi_s\xi)g(\phi_s\xi, \phi_s\xi) \\
&= -\varepsilon_{x_0}\varepsilon_{x_s}\varepsilon_\xi g(\phi_s\phi_i\phi_j\xi, \xi) \\
&= -\varepsilon_{x_0}\varepsilon_{x_s}\varepsilon_\eta g(\phi_s\phi_i\phi_j\eta, \eta),
\end{aligned}$$

and therefore they are independent of the unit vector ξ. Thus, we write (5.21) as

$$\phi_i\phi_j\xi = \sum_{s=1,\, s\neq i,j}^{\tau} \alpha_{ij}^s\phi_s\xi \tag{5.22}$$

for all unit vectors $\xi \in E_\lambda(y_0)$. Now, if we choose $k \in \{1,\ldots,\tau\}$ in a way that i,j,k are different, (5.22) leads to

$$\phi_i\phi_j(\phi_k\xi) = \sum_{s=1,\, s\neq i,j}^{\tau} \alpha_{ij}^s\phi_s(\phi_k\xi) = -\alpha_{ij}^k\varepsilon_{x_0}\varepsilon_{x_k}\xi + \sum_{s=1,\, s\neq i,j,k}^{\tau} \alpha_{ij}^s\phi_s\phi_k\xi.$$

On the other hand, (5.22) also gives

$$\begin{aligned}
\phi_k(\phi_i\phi_j\xi) = \phi_k\left(\sum_{s=1,\, s\neq i,j}^{\tau} \alpha_{ij}^s\phi_s\xi\right) &= \sum_{s=1,\, s\neq i,j}^{\tau} \alpha_{ij}^s\phi_k(\phi_s\xi) \\
&= -\alpha_{ij}^k\varepsilon_{x_0}\varepsilon_{x_k}\xi - \sum_{s=1,\, s\neq i,j,k}^{\tau} \alpha_{ij}^s\phi_s\phi_k\xi,
\end{aligned}$$

and since $\phi_i\phi_j(\phi_k\xi) = \phi_i\phi_j\phi_k\xi = \phi_k\phi_i\phi_j\xi = \phi_k(\phi_i\phi_j\xi)$, we obtain $\phi_i\phi_j\phi_k\xi = -\alpha_{ij}^k\varepsilon_{x_0}\varepsilon_{x_k}\xi$ for all unit vectors $\xi \in E_\lambda(y_0)$. This shows that the composition $\phi_i\phi_j$ coincides with ϕ_k or $-\phi_k$ on $E_\lambda(y_0)$ whenever i,j,k are different, in contradiction. Therefore, $dim\, M = 2(\tau + 1)$. $\square$

Next, we are concerned with the possible multiplicities of the eigenvalue λ. Note that the linear maps ϕ_i defined by (5.3.1) furnish each $E_\lambda(\cdot)$ with a Cliff$((\tau - 1)/2)$-module structure or a Cliff(τ)-module structure, depending on whether the signature of the restricted metric tensor to $E_\lambda(\cdot)$ is $(\tau+1,0)$, $(0,\tau+1)$ or $(\frac{\tau+1}{2}, \frac{\tau+1}{2})$. Now, we prove the following.

Theorem 5.3.1. *Let (M,g) be a semi-Riemannian special Osserman manifold. Then one of the following holds:*

(i) $\tau = 1$ and (M,g) is a 2n-dimensional manifold with metric tensor of signature (n,n) or $(2k,2r)$ for some $k,r \geq 0$,

(ii) $\tau = 3$ and (M,g) is a 4n-dimensional manifold with metric tensor of signature $(2n,2n)$ or $(4k,4r)$ for some $k,r \geq 0$,

(iii) $\tau = 7$ and (M,g) is a 16-dimensional manifold with metric tensor of signature $(8,8)$, $(16,0)$ or $(0,16)$,

(iv) $\tau = 15$ and (M,g) is a 32-dimensional manifold with metric tensor of signature $(16,16)$,

where τ denotes the multiplicity of the eigenvalue λ.

Proof. Let $y_0 \in S_p(M)$. By Remark 5.3.3, we can define $\tilde{\tau}$ complex structures $\phi_1, \ldots, \phi_{\tilde{\tau}}$ on $E_\lambda(y_0)$ which determine a $\mathrm{Cliff}(\tilde{\tau})$-module structure on $E_\lambda(y_0)$. Further note that $\dim E_\lambda(y_0) = \tilde{\tau} + 1$ if the restriction of the metric tensor to $E_\lambda(y_0)$ is of signature $(\tau+1, 0)$ or $(0, \tau+1)$. Moreover, $\dim E_\lambda(y_0) = 2\tilde{\tau} + 1$ if the induced inner product on $E_\lambda(y_0)$ is of signature $(\frac{\tau+1}{2}, \frac{\tau+1}{2})$.

Note that a $\mathrm{Cliff}(\tau)$-structure is available on a $(\tau+1)$-dimensional vector space if an only if the τ-dimensional sphere is parallelizable. This restricts us to the cases of $\tau = 1$, 3 and 7.

Now, if $\tilde{\tau} = \frac{\tau-1}{2}$, Theorem 2.1.1 shows that there exists such $\mathrm{Cliff}(\tilde{\tau})$-module structure if and only if

$$\tilde{\tau} \le \nu(r), \tag{5.23}$$

where $2\tilde{\tau} + 2 = 2^r \cdot n_0$, with odd n_0.

First, suppose that $\tilde{\tau}$ is even ($\tilde{\tau} = 2\alpha$). Then $2\tilde{\tau} + 2 = 2^r \cdot n_0$ is given by $2\tilde{\tau} + 2 = 4\alpha + 2 = 2(2\alpha + 1)$, and thus $r = 1$ and $n_0 = 2\alpha + 1$. Therefore, from (5.23), $2\alpha \le \nu(1) = 1$ and hence $\tilde{\tau}$ must be zero or odd.

Now suppose that $\tilde{\tau}$ is odd and it can be written in the form $\tilde{\tau} = 2^\alpha - 1$ for some α. In this case, $2\tilde{\tau} + 2 = 2^r \cdot n_0$ is given by $2\tilde{\tau} + 2 = 2(2^\alpha - 1) + 2 = 2^{\alpha+1}$ and hence $r = \alpha + 1$ and $n_0 = 1$. Thus, (5.23) shows that there exists a $\mathrm{Cliff}(\tilde{\tau})$-module structure on $E_\lambda(y_0)$ if and only if

$$2^\alpha - 1 \le \nu(\alpha + 1). \tag{5.24}$$

Now we consider this inequality. By putting $\alpha+1 = 4a+b$, $0 \le b \le 3$, we have $\nu(\alpha+1) = \nu(b) + 8a$ and since $\nu(b) = 2^b - 1$ and $a = (\alpha+1-b)/4$, it follows that $\nu(\alpha+1) = 2^b - 2b + 2\alpha + 1$. Then, (5.24) reduces to $2^\alpha - 1 \le 2^b - 2b + 2\alpha + 1$, that is, $2^\alpha - 2\alpha \le 2^b - 2b + 2$. Now, since $0 \le b \le 3$, we have $2^b - 2b + 2 \le 4$ and therefore,

$$2^\alpha - 2\alpha \le 4. \tag{5.25}$$

Consider the function $f(x) = 2^x - 2x$ with $x \in \mathbb{R}$. Since this function is strictly increasing on $(2, +\infty)$ and $f(4) = 8$, $2^\alpha - 2\alpha \ge 8$ whenever $\alpha \ge 4$. Thus, (5.25) gives a contradiction whenever $\alpha \ge 4$ and hence, $\tilde{\tau}$ cannot be written in the form $2^\alpha - 1$ for $\alpha \ge 4$. Now suppose that $\tilde{\tau}$ satisfies

$$2^\alpha - 1 < \tilde{\tau} < 2^{\alpha+1} - 1, \quad \alpha \ge 4. \tag{5.26}$$

Writing $2\tilde{\tau} + 2 = 2^r \cdot n_0$, with odd n_0, we know that (5.23) holds and moreover (5.26) implies that $2^{\alpha+1} < 2^r \cdot n_0 < 2^{\alpha+2}$. Thus it follows that $r \le \alpha + 1$ and hence

$$\nu(r) \leq \nu(\alpha + 1). \tag{5.27}$$

On the other hand, since $\alpha \geq 4$, (5.24) does not hold, and hence

$$2^\alpha - 1 > \nu(\alpha + 1). \tag{5.28}$$

Now note that (5.23), (5.26), (5.27) and (5.28) give

$$\tilde{\tau} \leq \nu(r) \leq \nu(\alpha + 1) < 2^\alpha - 1 < \tilde{\tau},$$

which means that $\tilde{\tau}$ cannot satisfy (5.26). Therefore, $\tilde{\tau} \in \{0, 1, 3, 5, 7, 9, 11, 13\}$. Also, a direct calculation from (5.23) shows that $\tilde{\tau} \in \{0, 1, 3, 7\}$.

Finally, since $\tilde{\tau} \in \{0, 1, 3, 7\}$, we have that $\tau \in \{1, 3, 7, 15\}$ and the result follows from Propositions 5.3.1 and 5.3.2. □

Remark 5.3.4. Note that the results in subsection 5.3.1 remain valid for any algebraic curvature tensor $\tilde{F}$ satisfying (I) and (II). However, it is worth pointing out the existence of many algebraic curvature tensors satisfying (I) but not (II). Indeed, if V is a $4k$-dimensional vector space furnished with an inner product $\langle \, , \, \rangle$ and a quaternionic structure $\mathbb{Q}$, then define $\tilde{F} = -\frac{1}{3}(R^{J_1} + R^{J_2})$, where $\{J_1, J_2, J_3 = J_1 J_2\}$ is a canonical basis for the quaternionic structure $\mathbb{Q}$. Then $\tilde{F}$ is an algebraic curvature tensor on V whose Jacobi operators are diagonalizable with exactly two eigenvalues, $\lambda = 1$ (of multiplicity $\tau = 2$) and $\mu = 0$. But note here that (II) fails since for each unit $x \in V$, $E_\lambda(x) = span\{x, J_1 x, J_2 x\}$, but $E_1(J_1 x) = span\{x, J_1 x, J_3 x\}$.

5.3.2 Semi-Riemannian Special Osserman Manifolds of Dimensions Different from 16 and 32

Next we explicitly determine the form of the curvature tensor of a semi-Riemannian special Osserman manifold when the multiplicity of the eigenvalue λ is assumed to be different from 7 and 15. Our purpose is to show that the curvature tensor of a semi-Riemannian special Osserman manifold whose eigenvalue λ, has multiplicity different from 7 and 15 can be writen at each point as a linear combination of R^0 and certain R^J's defined in Examples 1.2.1, 1.2.2 and 1.2.3.

More precisely, we prove the following:

Theorem 5.3.2. *Let (M, g) be a semi-Riemannian special Osserman manifold and assume that the multiplicity of the eigenvalue λ different from 7 and 15. Then, at each point $p \in M$, one of the following holds:*

(i) There exists a complex structure J on $T_p M$ such that (g, J) defines a Hermitian structure on $T_p M$ and the curvature tensor R of (M, g) at p is given by

$$R = \mu R^0 - \frac{\lambda - \mu}{3} R^J.$$

(ii) *There exists a paracomplex structure J on T_pM such that (g, J) defines a para-Hermitian structure on T_pM and the curvature tensor R of (M, g) at p is given by*

$$R = \mu R^0 + \frac{\lambda - \mu}{3} R^J.$$

(iii) *There exists a quaternionic structure $\mathbb{Q}$ on T_pM such that $(g, \mathbb{Q})$ defines a quaternionic Hermitian structure on T_pM and the curvature tensor R of (M, g) at p is given by*

$$R = \mu R^0 - \frac{\lambda - \mu}{3} \sum_{i=1}^{3} R^{J_i},$$

where $\{J_1, J_2, J_3\}$ is a canonical basis for $\mathbb{Q}$.

(iv) *There exists a paraquaternionic structure $\mathbb{Q}$ on T_pM such that $(g, \mathbb{Q})$ defines a paraquaternionic Hermitian structure on T_pM and the curvature tensor R of (M, g) at p is given by*

$$R = \mu R^0 + \frac{\lambda - \mu}{3} \sum_{i=1}^{3} \sigma_i R^{J_i},$$

where $\{J_1, J_2, J_3\}$ is a canonical basis for $\mathbb{Q}$ and $J_i^2 = \sigma_i id$, $i = 1, 2, 3$.

In order to prove Theorem 5.3.2, we need some technical lemmas.

Lemma 5.3.6. *Let (M, g) be a semi-Riemannian special Osserman manifold. Let $x_0, \xi \in S_p(M)$ with $\xi \in E_\lambda(x_0)^\perp$ and choose nonzero a, b such that $ax_0 + b\xi$ is also a unit vector. Put*

$$u_i = x_i - \frac{b}{a}\varepsilon_{x_0}\phi_i\xi, \qquad i = 1, \dots, \tau,$$

$$v_i = \frac{b}{a}\varepsilon_\xi x_i + \phi_i\xi, \qquad i = 1, \dots, \tau,$$

where $\{x_1, \dots, x_\tau\}$ is an orthonormal basis for $\ker(R_{x_0} - \lambda\varepsilon_{x_0}id)$ and $\phi_1, \dots, \phi_\tau$ are defined by (5.18). Then

$$R_{u_i}v_j = \mu g(u_i, u_i)v_j, \qquad i, j = 1, \dots, \tau.$$

Proof. Immediate from Lemma 5.3.4 and (5.18). $\qquad\square$

Lemma 5.3.7. *Let $x_0 \in S_p(M)$. Then for any unit ξ orthogonal to $E_\lambda(x_0)$,*

$$R(x_i, x_j)\xi = -\frac{2}{3}(\lambda - \mu)\varepsilon_{x_0}\phi_i\phi_j\xi, \qquad i \neq j, \quad i, j \in \{1, \dots, \tau\},$$

where $\{x_1, \dots, x_\tau\}$ is an orthonormal basis for $\ker(R_{x_0} - \lambda\varepsilon_{x_0}id)$ and $\phi_1, \dots, \phi_\tau$ are defined by (5.18).

Proof. Take $i, j \in \{1, \ldots, \tau\}$, $i \neq j$. Then since ξ is a unit vector orthogonal to $E_\lambda(x_0)$, so is $\phi_j \xi$, and therefore, if a and b are nonzero such that $a x_0 + b \phi_j \xi$ is a unit vector, Lemma 5.3.6 implies that

$$R_{u_j} v_i = \mu g(u_j, u_j) v_i, \tag{5.29}$$

where $u_j = x_j + \dfrac{b}{a} \varepsilon_{x_j} \xi$ and $v_i = \dfrac{b}{a} \varepsilon_{x_0} \varepsilon_{x_j} \varepsilon_\xi x_i + \phi_i \phi_j \xi$. Then,

$$
\begin{aligned}
R_{u_j} v_i &= R \left(\frac{b}{a} \varepsilon_{x_0} \varepsilon_{x_j} \varepsilon_\xi x_i + \phi_i \phi_j \xi, x_j + \frac{b}{a} \varepsilon_{x_j} \xi \right) \left(x_j + \frac{b}{a} \varepsilon_{x_j} \xi \right) \\
&= \frac{b}{a} \varepsilon_{x_0} \varepsilon_{x_j} \varepsilon_\xi R(x_i, x_j) x_j + \frac{b^3}{a^3} \varepsilon_{x_0} \varepsilon_{x_j} \varepsilon_\xi R(x_i, \xi) \xi \\
&\quad + \frac{b^2}{a^2} \varepsilon_{x_0} \varepsilon_\xi (R(x_i, x_j)\xi + R(x_i, \xi)x_j) + R(\phi_i \phi_j \xi, x_j)x_j \\
&\quad + \frac{b^2}{a^2} R(\phi_i \phi_j \xi, \xi)\xi + \frac{b}{a} \varepsilon_{x_j} (R(\phi_i \phi_j \xi, x_j)\xi + R(\phi_i \phi_j \xi, \xi)x_j).
\end{aligned}
$$

Also from Lemmas 5.3.1 and 5.3.2,

$$
\begin{aligned}
R_{u_j} v_i &= \frac{b}{a} \varepsilon_{x_0} \varepsilon_\xi \lambda x_i + \frac{b^3}{a^3} \varepsilon_{x_0} \varepsilon_{x_j} \mu x_i + \frac{3}{2} \frac{b^2}{a^2} \varepsilon_{x_0} \varepsilon_\xi R(x_i, x_j)\xi \\
&\quad + \varepsilon_{x_j} \mu \phi_i \phi_j \xi + \frac{b^2}{a^2} \varepsilon_\xi \lambda \phi_i \phi_j \xi + \frac{3}{2} \frac{b}{a} \varepsilon_{x_j} R(\phi_i \phi_j \xi, \xi) x_j \\
&= \left(\frac{b}{a} \varepsilon_{x_0} \varepsilon_\xi \lambda + \frac{b^3}{a^3} \varepsilon_{x_0} \varepsilon_{x_j} \mu \right) x_i + \frac{3}{2} \frac{b}{a} \varepsilon_{x_j} R(\phi_i \phi_j \xi, \xi) x_j \\
&\quad + \left(\varepsilon_{x_j} \mu + \frac{b^2}{a^2} \varepsilon_\xi \lambda \right) \phi_i \phi_j \xi + \frac{3}{2} \frac{b^2}{a^2} \varepsilon_{x_0} \varepsilon_\xi R(x_i, x_j)\xi.
\end{aligned}
$$

Therefore, since $g(u_j, u_j) = \varepsilon_{x_j} + \dfrac{b^2}{a^2} \varepsilon_\xi$, (5.29) is equivalent to

$$
\begin{aligned}
&\left(\frac{b}{a} \varepsilon_{x_0} \varepsilon_\xi \lambda + \frac{b^3}{a^3} \varepsilon_{x_0} \varepsilon_{x_j} \mu \right) x_i + \frac{3}{2} \frac{b}{a} \varepsilon_{x_j} R(\phi_i \phi_j \xi, \xi) x_j \\
&\quad + \left(\varepsilon_{x_j} \mu + \frac{b^2}{a^2} \varepsilon_\xi \lambda \right) \phi_i \phi_j \xi + \frac{3}{2} \frac{b^2}{a^2} \varepsilon_{x_0} \varepsilon_\xi R(x_i, x_j)\xi \\
&= \left(\frac{b}{a} \varepsilon_{x_0} \varepsilon_\xi \mu + \frac{b^3}{a^3} \varepsilon_{x_0} \varepsilon_{x_j} \mu \right) x_i + \left(\varepsilon_{x_j} \mu + \frac{b^2}{a^2} \varepsilon_\xi \mu \right) \phi_i \phi_j \xi,
\end{aligned}
$$

and hence

$$
\begin{aligned}
&\frac{b^2}{a^2} \varepsilon_\xi (\lambda - \mu) \phi_i \phi_j \xi + \frac{3}{2} \frac{b^2}{a^2} \varepsilon_{x_0} \varepsilon_\xi R(x_i, x_j)\xi \\
&\qquad = -\frac{b}{a} \varepsilon_{x_0} \varepsilon_\xi (\lambda - \mu) x_i - \frac{3}{2} \frac{b}{a} \varepsilon_{x_j} R(\phi_i \phi_j \xi, \xi) x_j.
\end{aligned}
$$

Now, since $\phi_i \phi_j \xi$ and $R(x_i, x_j)\xi$ belong to $E_\lambda(x_0)^\perp$ and x_i, $R(\phi_i \phi_j \xi, \xi)x_j \in E_\lambda(x_0)$, the previous expression gives

$$\frac{b^2}{a^2}\varepsilon_\xi(\lambda - \mu)\phi_i \phi_j \xi + \frac{3}{2}\frac{b^2}{a^2}\varepsilon_{x_0}\varepsilon_\xi R(x_i, x_j)\xi = 0.$$

It follows that $\frac{3}{2}\varepsilon_{x_0}R(x_i, x_j)\xi = -(\lambda - \mu)\phi_i \phi_j \xi$. Thus the claim follows. $\square$

Now we prove the main result of this subsection.

Proof of Theorem 5.3.2. Since the multiplicity of λ is assumed to be different from 7 and 15, Theorem 5.3.1 shows that the only possible multiplicities are 1 or 3. Now we analyze each case separately.

(a) λ of multiplicity $\tau = 1$.

Let $x_0 \in S_p(M)$ and take x_1 to be a unit vector in $ker(R_{x_0} - \lambda\varepsilon_{x_0}id)$. Let $\phi : E_\lambda(x_0)^\perp \longrightarrow E_\lambda(x_0)^\perp$ be the linear map defined by (5.18) so that $\phi^2 = \sigma Id$, where $\sigma = -\varepsilon_{x_0}\varepsilon_{x_1}$. Now, if y_0 is a unit vector orthogonal to $E_\lambda(x_0)$, take $y_1 = -\varepsilon_{y_0}\phi y_0$ and define $\psi : E_\lambda(x_0) \longrightarrow E_\lambda(x_0)$ by $\psi = \frac{3}{2(\lambda - \mu)}R(y_0, y_1)$ in a way that $\psi^2 = -\varepsilon_{y_0}\varepsilon_{y_1}Id = \sigma Id$. Now define $J : T_pM \longrightarrow T_pM$ by $J = \psi \oplus \phi$. Then J defines a complex (resp., paracomplex) structure on T_pM if $\sigma = -1$ (resp., $\sigma = +1$.)

Now we show that the curvature tensor R at $p \in M$ can be expressed in terms of the algebraic curvature tensors R^0 and R^J, where R^J is defined by the Hermitian or para-Hermitian structure (g, J). For this, we show that $\tilde{F} = R - \mu R^0 - \frac{\lambda - \mu}{3}\sigma R^J$ vanishes on T_pM.

First, we study the restriction of the trilinear map $\tilde{F}$ to the subspace $E_\lambda(x_0)$. It is easy to check that $\tilde{F}(E_\lambda(x_0), E_\lambda(x_0))E_\lambda(x_0) \subset E_\lambda(x_0)$. Thus, if we denote by $\bar{F}, \bar{R}, \bar{R}^0, \bar{R}^J$ the restrictions of $\tilde{F}, R, R^0, R^J$ to $E_\lambda(x_0)$, one has

$$\bar{R}_x = \lambda\varepsilon_x id_1, \quad \bar{R}_x^0 = \varepsilon_x id_1, \quad \bar{R}_x^J = 3\sigma\varepsilon_x id_1,$$

for all unit $x \in E_\lambda(x_0)$. Then $\bar{F}_x = 0$ and hence $\tilde{F}$ vanishes on $E_\lambda(x_0)$.

Now consider the restriction of $\tilde{F}$ to the subspace $E_\lambda(x_0)^\perp$. Note that this restriction is well-defined since $\tilde{F}(E_\lambda(x_0)^\perp, E_\lambda(x_0)^\perp)E_\lambda(x_0)^\perp \subset E_\lambda(x_0)^\perp$. Denote by $\hat{F}, \hat{R}, \hat{R}^0, \hat{R}^J$ the restrictions of $\tilde{F}, R, R^0, R^J$ to $E_\lambda(x_0)^\perp$. Now, if ξ is a unit vector in $E_\lambda(x_0)^\perp$ then the matrices of the associated Jacobi operators with respect to the orthonormal basis $\{J\xi, \eta_1, \ldots, \eta_{n-4}\}$ for the subspace $(span\{\xi\})^\perp \cap E_\lambda(x_0)^\perp$ become

$$\hat{R}_\xi = diag[\lambda\varepsilon_\xi, \mu\varepsilon_\xi, \overset{n-4}{\ldots}, \mu\varepsilon_\xi], \hat{R}_\xi^0 = diag[\varepsilon_\xi, \overset{n-3}{\ldots}, \varepsilon_\xi], \hat{R}_\xi^J = diag[3\sigma\varepsilon_\xi, 0, \overset{n-4}{\ldots}, 0],$$

and thus $\hat{F}_\xi = 0$. This shows that F is identically zero on $E_\lambda(x_0)^\perp$.

We proved that, considering the decomposition $T_pM = E_\lambda(x_0) \oplus E_\lambda(x_0)^\perp$, the algebraic curvature tensor $\tilde{F}$ vanishes when restricted to any of these subspaces. Thus, by Lemma 5.3.2, to show that $\tilde{F}$ is identically zero, it suffices to prove that $\tilde{F}(x_0, x_1) : E_\lambda(x_0)^\perp \longrightarrow E_\lambda(x_0)^\perp$ vanishes. For this, if ξ is a unit vector in $E_\lambda(x_0)^\perp$,

$$R(x_0, x_1)\xi = \frac{2}{3}(\lambda - \mu)J\xi, \quad \text{and} \quad R^0(x_0, x_1)\xi = 0. \quad (5.30)$$

Also,

$$g(Jx_0, x_1) = \frac{3}{2(\lambda - \mu)}g(R(y_0, y_1)x_0, x_1)$$

$$= \frac{3}{2(\lambda - \mu)}g(R(x_0, x_1)y_0, y_1)$$

$$= g(Jy_0, y_1),$$

and, since $y_1 = -\varepsilon_{y_0}Jy_0$, it follows that $g(Jx_0, x_1) = \sigma$. Thus

$$R^J(x_0, x_1)\xi = 2g(Jx_0, x_1)J\xi = 2\sigma J\xi. \quad (5.31)$$

Now, (5.30) and (5.31) give

$$\tilde{F}(x_0, x_1)\xi = \frac{2}{3}(\lambda - \mu)J\xi - \frac{\lambda - \mu}{3}\sigma(2\sigma J\xi) = 0,$$

which proves Theorem 5.3.2 when the multiplicity of λ is equal to 1.

(b) λ of multiplicity $\tau = 3$.

Let $x_0 \in S_p(M)$ and let $\{x_1, x_2, x_3\}$ be an orthonormal basis for $ker(R_{x_0} - \lambda\varepsilon_{x_0}id)$. Define $\phi_i : E_\lambda(x_0)^\perp \longrightarrow E_\lambda(x_0)^\perp$ by (5.18) such that $\phi_i^2 = \sigma_i id$, where $\sigma_i = -\varepsilon_{x_0}\varepsilon_{x_i}$, $i = 1, 2, 3$. Now, let y_0 be a unit vector orthogonal to $E_\lambda(x_0)$ and let $y_i = -\varepsilon_{y_0}\phi_i y_0$. Then $\{y_0, y_1, y_2, y_3\}$ determines an orthonormal basis for $E_\lambda(y_0)$ and this enables us to define the linear maps $\psi_i : E_\lambda(x_0) \longrightarrow E_\lambda(x_0)$ by $\psi_i = \frac{3}{2(\lambda - \mu)}R(y_0, y_i)$, such that $\psi_i^2 = -\varepsilon_{y_0}\varepsilon_{y_i}Id = \sigma_i Id$. (Note that, according to Remark 5.3.3, either $\sigma_1 = \sigma_2 = \sigma_3 = -1$ or $\sigma_1 = -\sigma_2 = -\sigma_3 = -1$.) For $i = 1, 2, 3$, let $J_i : T_pM \longrightarrow T_pM$ be defined by $J_i = \psi_i \oplus \phi_i$. Then J_i determines a complex or paracomplex structure on T_pM depending on whether σ_i is -1 or $+1$, respectively.

First we show that we may assume $J_1 J_2 = J_3$ without loss of generality. For this, consider the decomposition $T_pM = E_\lambda(x_0) \oplus E_\lambda(x_0)^\perp$. If ξ is a unit vector in any of these subspaces then $J_1 J_2\xi$ belongs to $E_\lambda(\xi)$ and, since $\{\xi, J_1\xi, J_2\xi, J_3\xi\}$ is an orthonormal basis and $J_1 J_2\xi \in (span\{\xi, J_1\xi, J_2\xi\})^\perp$, it follows that $J_1 J_2\xi \in span\{J_3\xi\}$. This proves that $J_1 J_2 = \pm J_3$ and, changing x_0 by $-x_0$ if necessary, we obtain $J_1 J_2 = J_3$.

The condition $J_1 J_2 = J_3$ also shows that $J_\alpha J_\beta = J_{\alpha\beta}$ for $\alpha, \beta = 1, 2, 3$, where the product $\alpha \cdot \beta$ is equal to the value $\pm\gamma$ such that $\pm e_\gamma = e_\alpha e_\beta$, where $\{e_0, e_1, e_2, e_3\}$ is the standard basis for the multiplication given by the following table:

	e_0	e_1	e_2	e_3
e_0	e_0	e_1	e_2	e_3
e_1	e_1	$\sigma_1 e_0$	e_3	$\sigma_1 e_2$
e_2	e_2	$-e_3$	$\sigma_2 e_0$	$-\sigma_2 e_1$
e_3	e_3	$-\sigma_1 e_2$	$\sigma_2 e_1$	$\sigma_3 e_0$

Note that this product corresponds to the quaternionic one if $\sigma_1 = \sigma_2 = \sigma_3 = -1$, and to the paraquaternionic one if $\sigma_1 = -\sigma_2 = -\sigma_3 = -1$. Also, put $J_0 = id$ and denote by $J_{-\alpha}$ the map $-J_\alpha$.

Next we show that the curvature tensor R at $p \in M$ can be expressed as in Theorem 5.3.2 by showing that $\tilde{F} = R - \mu R^0 - \dfrac{\lambda - \mu}{3} \sum_{i=1}^{3} \sigma_i R^{J_i}$ vanishes identically on $T_p M$. Again considering the orthogonal decomposition $T_p M = E_\lambda(x_0) \oplus E_\lambda(x_0)^\perp$ and proceeding the same way as in the case of multiplicity 1, we obtain that the algebraic curvature tensor $\tilde{F}$ vanishes when restricted to either of the subspaces $E_\lambda(x_0)$ or $E_\lambda(x_0)^\perp$. Therefore, it suffices to show that, for each $\alpha, \beta \in \{0, 1, 2, 3, 4\}$, and $\alpha < \beta$, $\tilde{F}(x_\alpha, x_\beta) : E_\lambda(x_0)^\perp \longrightarrow E_\lambda(x_0)^\perp$ vanishes identically.

First we consider $\tilde{F}(x_0, x_\alpha)$ for $\alpha = 1, 2, 3$. If ξ is a unit vector in $E_\lambda(x_0)^\perp$,

$$R(x_0, x_\alpha)\xi = \frac{2}{3}(\lambda - \mu)J_\alpha \xi, \quad R^0(x_0, x_\alpha)\xi = 0, \tag{5.32}$$

and

$$g(J_i x_0, x_\alpha) = \frac{3}{2(\lambda - \mu)} g(R(y_0, y_i)x_0, x_\alpha)$$

$$= \frac{3}{2(\lambda - \mu)} g(R(x_0, x_\alpha)y_0, y_i)$$

$$= g(J_\alpha y_0, y_i).$$

Since $y_i = -\varepsilon_{y_0} J_i y_0$, it follows that $g(J_i x_0, x_\alpha) = \delta_{i\alpha}\sigma_\alpha$ and thus,

$$R^{J_i}(x_0, x_\alpha)\xi = 2g(J_i x_0, x_\alpha)J_i \xi = 2\delta_{i\alpha}\sigma_\alpha J_\alpha \xi. \tag{5.33}$$

Now, (5.32) and (5.33) give

$$\tilde{F}(x_0, x_\alpha)\xi = \frac{2}{3}(\lambda - \mu)J_\alpha \xi - \frac{\lambda - \mu}{3} \sum_{i=1}^{3} \sigma_i(2\delta_{i\alpha}\sigma_\alpha J_\alpha \xi)$$

$$= \frac{2}{3}(\lambda - \mu)J_\alpha \xi - \frac{\lambda - \mu}{3} 2 J_\alpha \xi,$$

proving that $\tilde{F}(x_0, x_\alpha)$ vanishes.

Finally consider $F(x_\alpha, x_\beta)$ for $\alpha, \beta \in \{1, 2, 3\}$ and $\alpha < \beta$. If ξ is a unit vector in $E_\lambda(x_0)^\perp$, by $J_\alpha J_\beta = J_{\alpha\beta}$ and Lemma 5.3.7, we obtain

$$R(x_\alpha, x_\beta)\xi = -\frac{2}{3}(\lambda - \mu)\varepsilon_{x_0}J_{\alpha\beta}\xi, \qquad (5.34)$$

and

$$g(J_i x_\alpha, x_\beta) = \frac{3}{2(\lambda - \mu)}g(R(y_0, y_i)x_\alpha, x_\beta)$$

$$= \frac{3}{2(\lambda - \mu)}g(R(x_\alpha, x_\beta)y_0, y_i)$$

$$= -\varepsilon_{x_0}g(J_{\alpha\beta}y_0, y_i).$$

Therefore, since $y_i = -\varepsilon_{y_0}J_i y_0$, the previous expression gives $g(J_i x_\alpha, x_\beta)$ $= -\delta_{i,\alpha\beta}\sigma_{\alpha\beta}\varepsilon_{x_0}$, and hence

$$R^{J_i}(x_\alpha, x_\beta)\xi = 2g(J_i x_\alpha, x_\beta)J_i\xi = -2\delta_{i,\alpha\beta}\sigma_{\alpha\beta}\varepsilon_{x_0}J_{\alpha\beta}\xi. \qquad (5.35)$$

Now, since $R^0(x_\alpha, x_\beta)\xi = 0$, (5.34) and (5.35) imply that

$$\tilde{F}(x_\alpha, x_\beta)\xi = -\frac{2}{3}(\lambda - \mu)\varepsilon_{x_0}J_{\alpha\beta}\xi - \frac{\lambda - \mu}{3}\sum_{i=1}^{3}\sigma_i(-2\delta_{i,\alpha\beta}\sigma_{\alpha\beta}\varepsilon_{x_0}J_{\alpha\beta}\xi)$$

$$= -\frac{2}{3}(\lambda - \mu)\varepsilon_{x_0}J_{\alpha\beta}\xi + \frac{\lambda - \mu}{3}2\varepsilon_{x_0}J_{\alpha\beta}\xi = 0.$$

This completes the proof of Theorem 5.3.2. □

As an application of Theorem 5.3.2, we obtain a local version of Theorem 5.2.1 in the cases corresponding to λ of multiplicity 1 or 3. Therefore, now we can assume (M, g) is a semi-Riemannian special Osserman manifold whose curvature tensor is locally given by

$$R = \mu R^0 + \frac{\lambda - \mu}{3}\sum_{i=1}^{\tau}\sigma_i R^{J_i}, \qquad (5.36)$$

where $\tau = 1$, and (g, J_i) defines an indefinite almost Hermitian structure on a neigbordhood U_p of each $p \in M$ for $\sigma_i = -1$, or defines an almost para-Hermitian structure on U_p of each $p \in M$ for $\sigma_i = 1$. In the case $\tau = 3$, $(g, span\{J_1, J_2, J_3\})$ defines an indefinite quaternionic structure on U_p for $\sigma_1 = \sigma_2 = \sigma_3 = -1$, or a paraquaternionic structure on U_p for $\sigma_1 = -1, \sigma_2 = \sigma_3 = 1$. Further note that, since the curvature tensor R is given by (5.36), for each $\xi \in S_p(M)$, $E_\lambda(\xi) = span\{\xi, J_1\xi, \ldots, J_\tau\xi\}$.

First, we state several technical lemmas involving the covariant derivatives of J_i, which are needed in what follows.

Lemma 5.3.8. *Let* (M, g, J) *be an indefinite almost Hermitian or almost para-Hemitian manifold. Then*

(i) $(\nabla_X J)JY = -J(\nabla_X J)Y,$
(ii) $g((\nabla_X J)Y, Z) = -g(Y, (\nabla_X J)Z),$
(iii) $g((\nabla_X J)Y, Y) = g((\nabla_X J)Y, JY) = 0,$

for every $X, Y, Z \in \Gamma TM.$

Lemma 5.3.9. *Let* (M, g) *be a semi-Riemannian manifold whose curvature tensor* R *is given by* (5.36). *Then,*

$$(\nabla_X R)(Y, Z)T = \frac{\lambda - \mu}{3} \sum_{i=1}^{\tau} \sigma_i \{g(Y, J_i T)(\nabla_X J_i)Z + g(Y, (\nabla_X J_i)T)J_i Z$$

$$-g(Z, J_i T)(\nabla_X J_i)Y - g(Z, (\nabla_X J_i)T)J_i Y$$

$$+2g(Y, J_i Z)(\nabla_X J_i)T + 2g(Y, (\nabla_X J_i)Z)J_i DT\}$$

for every $X, Y, Z, T \in \Gamma TM.$

Lemma 5.3.10. *Let* (M, g) *be a semi-Riemannian manifold whose curvature tensor* R *is given by* (5.36). *Then,*

$$(\nabla_X J_s)X \in span\{J_i X, \, i \in \{1, \ldots, \tau\}, \, i \neq s\}, \qquad s = 1, \ldots, \tau,$$

for every unit $X \in \Gamma TM.$

Proof. Let Y be a unit vector field orthogonal to $E_\lambda(X)$. Then by the second Bianchi identity

$$(\nabla_Y R)(X, J_s Y)X + (\nabla_X R)(J_s Y, Y)X + (\nabla_{J_s Y} R)(Y, X)X = 0$$

and hence, it follows from Lemma 5.3.9 that

$$0 = \sum_{i=1}^{\tau} \sigma_i \{g(X, (\nabla_Y J_i)X)g(J_i J_s Y, Y) - g(Y, (\nabla_X J_i)X)g(J_i J_s Y, Y)$$

$$+2g(J_s Y, J_i Y)g((\nabla_X J_i)X, Y)\}.$$

Now, since $g(J_i Y, J_s Y) = -\delta_{is}\sigma_s \varepsilon_Y$ and $g(J_i J_s Y, Y) = -g(J_i Y, J_s Y)$, we have

$$0 = \sigma_s \{\sigma_s \varepsilon_Y g(X, (\nabla_Y J_s)X) - \sigma_s \varepsilon_Y g(Y, (\nabla_X J_s)X)$$

$$-2\sigma_s \varepsilon_Y g((\nabla_X J_s)X, Y)\}$$

$$= \varepsilon_Y \{g((\nabla_Y J_s)X, X) - 3g((\nabla_X J_s)X, Y)\}.$$

Thus Lemma 5.3.8-*(iii)* shows that $g((\nabla_X J_s)X, Y) = 0$. Also since Y is an arbitrary unit vector field in $E_\lambda(X)^\perp$, it follows that $(\nabla_X J_s)X \in E_\lambda(X)$ and the claim follows from Lemma 5.3.8. $\qquad\square$

Now we are ready to prove the following.

Theorem 5.3.3. *Let (M, g) be a semi-Riemannian special Osserman manifold with eigenvalue λ of multiplicity different from 7 and 15. Then, it is a locally symmetric space.*

Proof. Since the multiplicity of λ is assumed to be different from 7 and 15, the curvature tensor of (M, g) is given by (5.36) in a neighborhood of each point. Now we show that any semi-Riemannian manifold with curvature tensor R given by (5.36) is locally symmetric. Let X_0 be a unit vector field defined on an open subset $\mathcal{U}$ of M and consider the local decomposition

$$T\mathcal{U} = E_\lambda(X_0) \oplus E_\lambda(Y_0) \oplus \cdots.$$

We will show that

$$(\nabla_{X_0} R)(T, X_0, X_0, W) = 0, \tag{5.37}$$

for all vector fields T, W in an orthonormal frame induced by the decomposition above. First note that, from Lemma 5.3.9,

$$(\nabla_{X_0} R)(T, X_0, X_0, W)$$

$$= \frac{\lambda - \mu}{3} \sum_{i=1}^{\tau} \sigma_i \left\{ g(T, J_i X_0) g((\nabla_{X_0} J_i) X_0, W) + g(T, (\nabla_{X_0} J_i) X_0) g(J_i X_0, W) \right.$$

$$- g(X_0, J_i X_0) g((\nabla_{X_0} J_i) T, W) - g(X_0, (\nabla_{X_0} J_i) X_0) g(J_i T, W)$$

$$\left. + 2g(T, J_i X_0) g((\nabla_{X_0} J_i) X_0, W) + 2g(T, (\nabla_{X_0} J_i) X_0) g(J_i X_0, W) \right\},$$

and since X_0 is orthogonal to $J_i X_0$ and $(\nabla_{X_0} J_i) X_0$,

$$(\nabla_{X_0} R)(T, X_0, X_0, W) = (\lambda - \mu) \sum_{i=1}^{\tau} \sigma_i \left\{ g(T, J_i X_0) g(W, (\nabla_{X_0} J_i) X_0) \right.$$

$$\left. + g(T, (\nabla_{X_0} J_i) X_0) g(W, J_i X_0) \right\}.$$

Now, by using Lemma 5.3.10, note that the previous expression vanishes whenever either of the vector fields T and W belongs to $\Gamma E_\lambda(X_0)^\perp$, and thus, (5.37) holds for such choices of T and W.

To complete the proof, we analyze the case of T, $W \in \Gamma E_\lambda(X_0)$. Note that

$$(\nabla_{X_0} R)(T, X_0, X_0, W) = \nabla_{X_0}(R(T, X_0, X_0, W)) - R(\nabla_{X_0} T, X_0, X_0, W)$$

$$- R(T, \nabla_{X_0} X_0, X_0, W) - R(T, X_0, \nabla_{X_0} X_0, W)$$

$$- R(T, X_0, X_0, \nabla_{X_0} W).$$

But since

$$\nabla_{X_0}(R(T, X_0, X_0, W)) - R(\nabla_{X_0}T, X_0, X_0, W) - R(T, X_0, X_0, \nabla_{X_0}W)$$

$$= \nabla_{X_0}(g(R_{X_0}T, W)) - g(R_{X_0}W, \nabla_{X_0}T) - g(R_{X_0}T, \nabla_{X_0}W)$$

$$= \nabla_{X_0}(\lambda \varepsilon_{X_0} g(T, W)) - \lambda \varepsilon_{X_0} g(W, \nabla_{X_0}T) - \lambda \varepsilon_{X_0} g(T, \nabla_{X_0}W)$$

$$= \lambda \varepsilon_{X_0} \{\nabla_{X_0}(g(T, W)) - g(\nabla_{X_0}T, W) - g(T, \nabla_{X_0}W)\}$$

$$= \lambda \varepsilon_{X_0} (\nabla_{X_0}g)(T, W)$$

$$= 0,$$

it follows that

$$(\nabla_{X_0}R)(T, X_0, X_0, W) = -R(T, \nabla_{X_0}X_0, X_0, W) - R(T, X_0, \nabla_{X_0}X_0, W). \tag{5.38}$$

Now, put $T = X_i$ and $W = X_j$, with $i, j \in \{1, \dots, \tau\}$. If $i \neq j$, Lemma 5.3.2-(iii) shows that (5.38) vanishes. If $i = j$, from (5.38), we obtain

$$(\nabla_{X_0}R)(X_i, X_0, X_0, X_i) = -R(X_i, \nabla_{X_0}X_0, X_0, X_i) - R(X_i, X_0, \nabla_{X_0}X_0, X_i)$$

$$= -2R(X_0, X_i, X_i, \nabla_{X_0}X_0)$$

$$= -2g(R_{X_i}X_0, \nabla_{X_0}X_0)$$

$$= -2\lambda \varepsilon_{X_i} g(X_0, \nabla_{X_0}X_0),$$

which also vanishes since $g(X_0, \nabla_{X_0}X_0) = 0$. This shows that (5.37) holds and therefore, (M, g) is locally symmetric. $\qquad\square$

As pointed out in subsection 5.3.2, our main purpose in this subsection is to prove a local version of Theorem 5.2.1 for the characterization of semi-Riemannian special Osserman manifolds. Now, we state and prove it for the special case when the multiplicity of λ is different from 7 and 15.

Theorem 5.3.4. *Let (M, g) be a semi-Riemannian special Osserman manifold. If the multiplicity of the distinguished eigenvalue λ is different from 7 or 15 then (M, g) is locally isometric to one of the following:*

(a) an indefinite complex space form with metric tensor of signature $(2k, 2r)$, $k, r \geq 0$,

(b) an indefinite quaternionic space form with metric tensor of signature $(4k, 4r)$, $k, r \geq 0$,

(c) a paracomplex space form with metric tensor of signature (k, k),

(d) a paraquaternionic space form with metric tensor of signature $(2k, 2k)$.

Proof. Since the multiplicity of λ is different from 7 and 15, (M, g) is locally symmetric and its curvature tensor R is given by (5.36). Also note that, as an immediate consequence of the definition of R^{J_i}'s, one has

$$\left(\nabla_X R^{J_i}\right)(Y, Z)Z = 3\left\{g((\nabla_X J_i)Z, Y)J_i Z + g(Y, J_i Z)(\nabla_X J_i)Z\right\}, \tag{5.39}$$

for every $X, Y, Z \in \Gamma TM$.

Now, we start to analyze separately the following four possible cases.

(a): $\tau = 1$ and (g, J) defines an indefinite almost Hermitian structure.

Since (M, g) is locally symmetric, (5.36) implies that $\nabla R^J = 0$ and, in particular, $(\nabla_X R^J)(JY, Y)Y = 0$ for all unit vector fields X, Y on M. Now, (5.39) and Lemma 5.3.8 give $(\nabla_X R^J)(JY, Y)Y = 3\varepsilon_Y(\nabla_X J)Y$, and hence $(\nabla_X J)Y = 0$. This shows that (g, J) is an indefinite Kähler structure.

On the other hand, if $x, y \in T_p M$ are nonnull with $y \in (span\{x, Jx\})^\perp$, since $R^0(x, Jx, x, Jy) = R^J(x, Jx, x, Jy) = 0$, (5.36) implies that $R(x, Jx, x, Jy) = 0$. Hence, the holomorphic sectional curvature is constant (cf. [4, Thm 5.1].)

(b): $\tau = 1$ and (g, J) defines an almost para-Hermitian structure.

Proceeding in the same way above, it follows that (g, J) is a para-Kähler structure. Moreover, it also follows from (5.36) that $R(x, Jx, x, Jy) = 0$ for every $x, y \in T_p M$ with $y \in (span\{x, Jx\})^\perp$. Hence, the paraholomorphic sectional curvature is constant (cf. [54, Thm.].)

(c): $\tau = 3$ and $(g, span\{J_1, J_2, J_3\})$ defines an indefinite quaternionic structure.

Since M is locally symmetric and the algebraic curvature tensor R^0 is parallel, it follows from (5.36) that $\sum_{i=1}^{3} (\nabla_X R^{J_i})(Z, Y)Y = 0$ for every $X, Y, Z \in \Gamma TM$. Now, if $Z \in \Gamma E_\lambda(Y)^\perp$ then, (5.39) and Lemma 5.3.3 give $(\nabla_X R^{J_i})(Z, Y)Y = 3g((\nabla_X J_i)Y, Z)J_i Y$, and hence $g((\nabla_X J_i)Y, Z) = 0$ whenever $Z \in \Gamma E_\lambda(Y)^\perp$. This shows that $V = span\{J_1, J_2, J_3\}$ is parallel and therefore the indefinite quaternionic structure is Kähler.

Now, if $x, y \in T_p M$ are nonnull with $y \in (span\{x, J_1 x, J_2 x, J_3 x\})^\perp$, it follows from (5.36) that $R(x, J_i x, x, J_i y) = 0$ for all $i = 1, 2, 3$ and therefore, the quaternionic sectional curvature is constant (cf. [118, Lemma 5.4].)

(d): $\tau = 3$ and $(g, span\{J_1, J_2, J_3\})$ defines a paraquaternionic structure.

Similar to the case (c), we prove that the paraquaternionic structure is Kähler. Moreover, if $x, y \in T_p M$ are nonnull with $y \in (span\{x, J_1 x, J_2 x, J_3 x\})^\perp$, then as in (c), we obtain $R(x, J_i x, x, J_i y) = 0$, $i = 1, 2, 3$, and therefore the paraquaternionic sectional curvature is constant (cf. [62, Thm 4.1].) $\square$

5.3.3 Semi-Riemannian Special Osserman Manifolds of Dimensions 16 and 32

The purpose of this subsection is to prove the remaining part of Theorem 5.2.1. That is, to study further the cases when the multiplicity of λ is equal to 7 (dim $M = 16$) or 15 (dim $M = 32$.)

Let X, Y be unit vectors fields with $Y \in E_\lambda(X)^\perp$ and let $\{X_1, \ldots, X_\tau\}$ be an orthonormal basis for $ker(R_X - \lambda \varepsilon_X id)$. Further, let $\{Y_0 = Y, Y_1 = \phi_1 Y_0, \ldots, Y_\tau = \phi_\tau Y_0\}$ be an orthonormal basis for $E_\lambda(Y)$, where the linear maps ϕ_i are defined by $\phi_i = \frac{3}{2(\lambda-\mu)} R(X_0, X_i)$ (see (5.18).) Now, define a product on $E_\lambda(Y)$ by

$$Y_0 \cdot Y_i = Y_i \cdot Y_0 = Y_i, \qquad Y_i \cdot Y_j = \phi_i \phi_j Y_0, \qquad i,j = 1, \ldots, \tau.$$

In the following let $\{e_0, e_1, \ldots, e_\tau\}$ denote the standard basis for the product above, that is, a basis for $E_\lambda(Y)$ such that $e_\alpha \cdot e_\beta$ is a basic element, and write $e_{\alpha\beta} = e_\alpha \cdot e_\beta$.

In the case $\tau = 7$, such a product is given by the following table:

	e_0	e_1	e_2	e_3	e_4	e_5	e_6	e_7
e_0	e_0	e_1	e_2	e_3	e_4	e_5	e_6	e_7
e_1	e_1	$-e_0$	e_3	$-e_2$	$-e_5$	e_4	$-e_7$	e_6
e_2	e_2	$-e_3$	$-e_0$	e_1	$-e_6$	e_7	e_4	$-e_5$
e_3	e_3	e_2	$-e_1$	$-e_0$	$-e_7$	$-e_6$	e_5	e_4
e_4	e_4	e_5	e_6	e_7	$-\varepsilon e_0$	$-\varepsilon e_1$	$-\varepsilon e_2$	$-\varepsilon e_3$
e_5	e_5	$-e_4$	$-e_7$	e_6	εe_1	$-\varepsilon e_0$	$-\varepsilon e_3$	εe_2
e_6	e_6	e_7	$-e_4$	$-e_5$	εe_2	εe_3	$-\varepsilon e_0$	$-\varepsilon e_1$
e_7	e_7	$-e_6$	e_5	$-e_4$	εe_3	$-\varepsilon e_2$	εe_1	$-\varepsilon e_0$

Now, it is easy to see that it corresponds to the product of the octonians (if $\varepsilon = \varepsilon_{Y_0} \varepsilon_W = 1$) and the product of the anti-octonians (if $\varepsilon = \varepsilon_{Y_0} \varepsilon_W = -1$.) (See [102])

Now, we have

Lemma 5.3.11. *Let (M, g) be a semi-Riemannian special Osserman manifold with the eigenvalue λ of multiplicity $\tau = 7$ or 15, and let X, Y be unit vector fields on M with $Y \in E_\lambda(X)^\perp$. Then, there exists an orthonormal basis $\{X_1, \ldots, X_\tau\}$ for $ker(R_X - \lambda \varepsilon_X id)$ such that*

$$\phi_\alpha \phi_\beta Y_0 = \phi_{\alpha\beta} Y_0, \qquad \alpha, \beta \in \{1, \ldots, \tau\}$$

where $\phi_0 = Id$ and $\phi_{-\alpha} = -\phi_\alpha$.

Proof. Let $\{X_1, \ldots, X_\tau\}$ be an orthonormal basis for $ker(R_{X_0} - \lambda \varepsilon_{X_0} id)$ and consider the associated bundle homomorphisms ϕ_i defined by (5.18) at each point. Moreover let $\{Y_0, Y_1, \ldots, Y_\tau\}$ denote the induced basis for $E_\lambda(Y_0)$ given by $Y_i = \phi_i Y_0$, $i = 1, \ldots, \tau$.

Now if $\{e_0 = Y_0, e_1, \ldots, e_\tau\}$ is the standard basis for the product in $E_\lambda(Y_0)$, that is $e_\alpha \cdot e_\beta = e_{\alpha\beta}$, then express the elements e_i with respect to the basis $\{Y_0, \ldots, Y_\tau\}$ by

$$e_0 = Y_0, \qquad e_i = \sum_{j=0}^{\tau} a_{ij} Y_j, \qquad i = 1, \ldots, \tau,$$

and define

$$\overline{X}_0 = X_0, \qquad \overline{X}_i = \sum_{j=0}^{\tau} a_{ij} X_j, \qquad i = 1, \ldots, \tau,$$

in $E_\lambda(X_0)$. Now, we show that $\{\overline{X}_0, \overline{X}_1, \ldots, \overline{X}_\tau\}$ is the required basis. First note that, if $\overline{\phi}_1, \ldots, \overline{\phi}_\tau$ denote the associated bundle homomorphisms defined by such a basis, then

$$\overline{\phi}_\alpha \overline{\phi}_\beta Y_0 = \left\{ \tfrac{3}{2(\lambda-\mu)} \right\}^2 R(\overline{X}_0, \overline{X}_\alpha) R(\overline{X}_0, \overline{X}_\beta) Y_0$$

$$= \left\{ \tfrac{3}{2(\lambda-\mu)} \right\}^2 R\left(X_0, \textstyle\sum_{r=0}^{\tau} a_{\alpha r} X_r\right) R\left(X_0, \textstyle\sum_{s=0}^{\tau} a_{\beta s} X_s\right) Y_0$$

$$= \left\{ \tfrac{3}{2(\lambda-\mu)} \right\}^2 \sum_{r,s=1}^{\tau} a_{\alpha r} a_{\beta s} R(X_0, X_r) R(X_0, X_s) Y_0$$

$$= \left\{ \tfrac{3}{2(\lambda-\mu)} \right\}^2 \sum_{r,s=1}^{\tau} a_{\alpha r} a_{\beta s} R(X_0, X_r) \left(\frac{2(\lambda-\mu)}{3} \phi_s Y_0 \right)$$

$$= \tfrac{3}{2(\lambda-\mu)} \sum_{r,s=1}^{\tau} a_{\alpha r} a_{\beta s} R(X_0, X_r) \phi_s Y_0$$

$$= \frac{3}{2(\lambda-\mu)} \sum_{r,s=1}^{\tau} a_{\alpha r} a_{\beta s} \frac{2(\lambda-\mu)}{3} \phi_r \phi_s Y_0$$

$$= \sum_{r,s=1}^{\tau} a_{\alpha r} a_{\beta s} (Y_r \cdot Y_s) = e_\alpha \cdot e_\beta.$$

On the other hand,

$$\overline{\phi}_{\alpha\beta} Y_0 = \tfrac{2}{3(\lambda-\mu)} R(\overline{X}_0, \overline{X}_{\alpha\beta}) Y_0$$

$$= \tfrac{3}{2(\lambda-\mu)} R\left(X_0, \textstyle\sum_{t=0}^{\tau} a_{(\alpha\beta)t} Y_t\right) Y_0$$

$$= \frac{3}{2(\lambda-\mu)} \sum_{t=0}^{\tau} a_{(\alpha\beta)t} R(X_0, X_t) Y_0$$

$$= \frac{3}{2(\lambda-\mu)} \sum_{t=0}^{\tau} a_{(\alpha\beta)t} \frac{2(\lambda-\mu)}{3} \phi_t Y_0$$

$$= \sum_{t=0}^{\tau} a_{(\alpha\beta)t} Y_t = e_{\alpha\beta}.$$

Now the result follows from the previous expressions by using that $e_\alpha \cdot e_\beta = e_{\alpha\beta}$. □

Theorem 5.3.5. *Let (M, g) be a semi-Riemannian special Osserman manifold with the eigenvalue λ of multiplicity $\tau = 7$ or 15. Then it is locally symmetric.*

Proof. Let X be a unit vector field on M. We show that $(\nabla_X R)(T, X, X, W) = 0$ for every $W, T \in \Gamma TM$. For this, let Y be a unit vector field in $\Gamma E_\lambda(X)^\perp$ and let $\{X_1, \ldots, X_\tau\}$ be a basis for $ker(R_{X_0} - \lambda\varepsilon_{X_0} id)$ given by Lemma 5.3.11. Now we show that $(\nabla_X R)(T, X, X, W) = 0$ for every T, W taken in $\Gamma span\{X_0 = X, X_1, \ldots, X_\tau, Y_0 = Y, Y_1, \ldots, Y_\tau\}$, where $Y_i = \phi_i Y_0$, $i = 1, \ldots, \tau$.

First, if $T \neq W \in \Gamma E_\lambda(X)$ then

$$(\nabla_X R)(T, X, X, W) = \nabla_X R(T, X, X, W) - R(\nabla_X T, X, X, W)$$
$$-R(T, \nabla_X X, X, W) - R(T, X, \nabla_X X, W) - R(T, X, X, \nabla_X W)$$
$$= \nabla_X \lambda\varepsilon_X g(T, W) - \lambda\varepsilon_X g(W, \nabla_X T) - \lambda\varepsilon_X g(T, \nabla_X W)$$
$$-R(T, \nabla_X X, X, W) - R(T, X, \nabla_X X, W)$$
$$= \lambda\varepsilon_X (\nabla_X g)(T, W) = 0 \qquad \text{by Lemma 5.3.2-(iii).}$$

Also, if $T = W \in \Gamma E_\lambda(X)$ then we have

$$(\nabla_X R)(W, X, X, W) = \nabla_X R(W, X, X, W) - R(\nabla_X W, X, X, W)$$
$$-R(W, \nabla_X X, X, W) - R(W, X, \nabla_X X, W) - R(W, X, X, \nabla_X W)$$
$$= \lambda\varepsilon_X (\nabla_X g)(W, W) - 2R(X, W, W, \nabla_X X)$$
$$= -2\lambda\varepsilon_W g(X, \nabla_X X) = 0.$$

Similarly, if $T \neq W \in \Gamma E_\lambda(Y)$ then

$$(\nabla_X R)(T, X, X, W) = \nabla_X R(T, X, X, W) - R(\nabla_X T, X, X, W)$$
$$-R(T, \nabla_X X, X, W) - R(T, X, \nabla_X X, W) - R(T, X, X, \nabla_X W)$$
$$= \nabla_X \mu\varepsilon_X g(T, W) - \mu\varepsilon_X g(W, \nabla_X T) - \mu\varepsilon_X g(T, \nabla_X W)$$
$$-R(T, \nabla_X X, X, W) - R(T, X, \nabla_X X, W)$$
$$= \mu\varepsilon_X (\nabla_X g)(T, W) - R(X, W, T, \nabla_X X) + R(T, X, W, \nabla_X X)$$
$$= 0 \qquad \text{by Lemma 5.3.2-(ii).}$$

Lastly, if $T = W \in \Gamma E_\lambda(Y)$ then

$$(\nabla_X R)(W, X, X, W) = \nabla_X R(W, X, X, W) - R(\nabla_X W, X, X, W)$$
$$-R(W, \nabla_X X, X, W) - R(W, X, \nabla_X X, W) - R(W, X, X, \nabla_X W)$$
$$= \mu\varepsilon_X (\nabla_X g)(W, W) - 2R(X, W, W, \nabla_X X)$$
$$= -2\mu\varepsilon_W g(X, \nabla_X X) = 0.$$

To complete the proof, let us consider the case $T = X_i$, $W = Y_j$. Now, from Lemma 5.3.2-(ii) and Lemma 5.3.11,

$$(\nabla_X R)(X_i, X, X, Y_j) = \nabla_X R(X_i, X, X, Y_j) - R(\nabla_X X_i, X, X, Y_j)$$
$$-R(X_i, \nabla_X X, X, Y_j) - R(X_i, X, \nabla_X X, Y_j) - R(X_i, X, X, \nabla_X Y_j)$$
$$= (\lambda - \mu)\{\varepsilon_X g(Y_j, \nabla_X X_i) - g(Y_{ij}, \nabla_X X)\}.$$

Once again from Lemma 5.3.2-(ii),

$$(\nabla_X R)(Y_{ij}, Y_j, Y_j, X) = \nabla_X R(Y_{ij}, Y_j, Y_j, X) - R(\nabla_X Y_{ij}, Y_j, Y_j, X)$$
$$-R(Y_{ij}, \nabla_X Y_j, Y_j, X) - R(Y_{ij}, Y_j, \nabla_X Y_j, X) - R(Y_{ij}, Y_j, Y_j, \nabla_X X)$$
$$= \mu\varepsilon_{Y_j} g(Y_{ij}, \nabla_X X) - \lambda\varepsilon_{Y_j} g(Y_{ij}, \nabla_X X) - \tfrac{3}{2} R(Y_j, Y_{ij}, X, \nabla_X Y_j).$$

Now, by Lemma 5.3.2-(i), $R(Y_j, Y_{ij})X = \sum_{k=0}^{\tau} R(Y_j, Y_{ij}, X, X_k)\varepsilon_{X_k} X_k$ and hence, since $R(Y_j, Y_{ij}, X, X_k) = R(Y_j, Y_{ij}, X, X_k) = \sum_{k=0}^{\tau} \tfrac{2}{3}(\lambda-\mu)g(Y_{kj}, Y_{ij})$, it follows that $R(Y_j, Y_{ij})X = \tfrac{2}{3}(\lambda - \mu)\varepsilon_X \varepsilon_{Y_j} X_i$. Thus

$$(\nabla_X R)(Y_{ij}, Y_j, Y_j, X) = (\lambda - \mu)\varepsilon_{Y_j}\{\varepsilon_X g(Y_j, \nabla_X X_i) - g(Y_{ij}, \nabla_X X)\}$$

and therefore, $(\nabla_X R)(X_i, X, X, Y_j) = \varepsilon_{Y_j}(\nabla_X R)(Y_{ij}, Y_j, Y_j, X)$. Finally, by the second Bianchi identity,

$$(\nabla_X R)(Y_{ij}, Y_j, Y_j, X) = -(\nabla_{Y_{ij}} R)(Y_j, X, X, Y_j) - (\nabla_{Y_j} R)(Y_{ij}, X, X, Y_j)$$

and this expression vanishes as in the previous case of $T, W \in \Gamma E_\lambda(Y)$. Thus it follows that $(\nabla_X R)(X_i, X, X, Y_j) = 0$ and hence, (M, g) is locally symmetric.   □

proof of Theorem 5.2.1. The cases (a), (b), (c) and (d) in Theorem 5.2.1 follow immediately from Theorem 5.3.4. Therefore, we concentrate on the cases when the multiplicity of λ is 7 or 15.

First, if we consider a basis for the tangent space $T_p M$ given by Lemma 5.3.11 then, by a straightforward computation using (5.18) and Lemmas 5.3.2 and 5.3.7, we obtain the following expressions for the curvature operators $R(x, y)$.

For each $i = 1, \ldots, \tau$,

$$R(x_0, x_i)x_\alpha = \begin{cases} -\lambda\varepsilon_{x_0} x_i & \alpha = 0 \\ \lambda\varepsilon_{x_i} x_0 & \alpha = i \\ 0 & \alpha \neq 0, i \end{cases}$$

$$R(x_0, x_i)y_\alpha = \frac{2(\lambda - \mu)}{3} y_{i\alpha}$$

and

$$R(x_0, y_0)x_\alpha = \begin{cases} -\mu\varepsilon_{x_0}y_0 & \alpha = 0 \\ \frac{\lambda-\mu}{3}y_\alpha & \alpha \neq 0 \end{cases},$$

$$R(x_0, y_0)y_\alpha = \begin{cases} \mu\varepsilon_{y_0}x_0 & \alpha = 0 \\ -\frac{\lambda-\mu}{3}\varepsilon_{x_0}\varepsilon_{y_0}x_\alpha & \alpha \neq 0. \end{cases}$$

Moreover, for each $i, j = 1, \ldots, \tau$ with $i \neq j$, we have

$$R(x_i, x_j)x_\alpha = \begin{cases} -\lambda\varepsilon_{x_i}x_j & \alpha = i \\ \lambda\varepsilon_{x_j}x_i & \alpha = j \\ 0 & \alpha \neq i, j \end{cases}$$

$$R(x_i, x_j)y_\alpha = -\frac{2(\lambda-\mu)}{3}\varepsilon_{x_0}y_{i(j\alpha)}.$$

Also, for each $i = 1, \ldots, \tau$,

$$R(x_i, y_0)x_\alpha = \begin{cases} -\frac{\lambda-\mu}{3}y_i & \alpha = 0 \\ -\mu\varepsilon_{x_i}y_0 & \alpha = i \\ -\frac{\lambda-\mu}{3}\varepsilon_{x_0}y_{i\alpha} & \alpha \neq 0, i \end{cases}$$

$$R(x_i, y_0)y_\alpha = \begin{cases} \mu\varepsilon_{y_0}x_i & \alpha = 0 \\ \frac{\lambda-\mu}{3}\varepsilon_{x_0}\varepsilon_{y_i}x_0 & \alpha = i \\ -\frac{\lambda-\mu}{3}\varepsilon_{y_0}x_{i\alpha} & \alpha \neq 0, i. \end{cases}$$

Now, it follows from the previous expressions that the action of the holonomy group on each tangent space is irreducible and hence a complete, simply connected semi-Riemannian special Osserman manifold with the eigenvalue λ of multiplicity $\tau = 7$ or 15 must be one of the symmetric spaces in Berger's list [7, p. 157]. In order to complete the proof of Theorem 5.2.1, we consider separately the different possibilities corresponding to $\tau = 7$ and $\tau = 15$.

• **The 16-dimensional case; $\tau = 7$.**

By a straightforward computation, it can be shown that the curvature operators $R(x_i, x_j)$, $i, j = 0, \ldots, 7, i < j$ and $R(x_i, y_0)$, $i = 1, \ldots, 7$ are linearly independent. Thus, the dimension of the isotropy group of a semi-Riemannian special Osserman manifold with the eigenvalue λ of multiplicity 7 must be ≥ 36.

Moreover, the metric tensor of a semi-Riemannian special Osserman manifold with eigenvalue λ of multiplicity 7 must be either of signatures $(16, 0)$, $(0, 16)$ or $(8, 8)$ (cf. Theorem 5.3.1.) Thus, the only candidates in Berger's list are $\frac{SL(9,\mathbb{R})}{SL(8,\mathbb{R})+\mathbb{R}}$, $\frac{SO(9,\mathbb{C})}{SO(8,\mathbb{C})}$, $\frac{Sp(5,\mathbb{R})}{Sp(1,\mathbb{R})+Sp(4,\mathbb{R})}$, $\frac{F_4}{SO(9)}$, $\frac{F_4^2}{SO(9)}$, $\frac{F_4^2}{SO^1(9)}$, $\frac{F_4^1}{SO^4(9)}$, and

$$\frac{SU^i(n)}{SU^k(k+h)+SU^{i-k}(n-k-h)+T}, \quad (k+h)(n-k-h) = 8,$$

$$\frac{SO^i(n)}{SO^k(k+h)+SO^{i-k}(n-k-h)}, \quad \begin{aligned}(k+h)(n-k-h) &= 16, \\ k+h &> 2, n-k-h > 2,\end{aligned}$$

$$\frac{Sp^i(n)}{Sp^k(k+h)+Sp^{i-k}(n-k-h)+T}, \quad (k+h)(n-k-h) = 4.$$

Now, note that $\frac{SL(9,\mathbb{R})}{SL(8,\mathbb{R})+\mathbb{R}}$ and $\frac{Sp(5,\mathbb{R})}{Sp(1,\mathbb{R})+Sp(4,\mathbb{R})}$ correspond to the 16-dimensional paracomplex and paraquaternionic space forms, respectively (see [53], [62].) Moreover, $\frac{SO(9,\mathbb{C})}{SO(8,\mathbb{C})}$ corresponds to the complex sphere $\mathbb{C}S^8$ which can be viewed as a hypersurface in the indefinite sphere. It easily follows from [103] that such complex spheres are not Osserman manifolds. Now, it also follows that $\frac{SO^i(n)}{SO^k(k+h)+SO^{i-k}(n-k-h)}$ occurs only if $n = 8$, but it must be excluded since in this case the dimension of the holonomy group is < 36. The symmetric spaces $\frac{Sp^i(n)}{Sp^k(k+h)+Sp^{i-k}(n-k-h)+T}$ can only happen if $n = 5$ and they correspond to the indefinite quaternionic projective or hyperbolic spaces. Also, by an argument on the dimension of the holonomy group, $\frac{SU^i(n)}{SU^k(k+h)+SU^{i-k}(n-k-h)+T}$ may only occur if $n = 9$, but in this case, they correspond to the indefinite complex projective or hyperbolic spaces.

The remaining spaces, $F_4/SO(9)$, $F_4^2/SO(9)$, $F_4^2/SO^1(9)$ and $F_4^1/SO^4(9)$ correspond to the Cayley planes listed in Theorem 5.2.1 (see also [125].)

- **The 32-dimensional case; $\tau = 15$.**

Proceeding as in the previous case, the curvature operators $R(x_i, x_j)$, $i, j = 0, \ldots, 15, i < j$, and $R(x_i, y_0)$, $i = 1, \ldots, 15$, are linearly independent and hence, the dimension of the isotropy group of any semi-Riemannian special Osserman manifold with the eigenvalue λ of multiplicity 15 must be ≥ 136.

Moreover, since such a semi-Riemannian special Osserman manifold must have a metric tensor of signature $(16, 16)$ (cf. Theorem 5.3.1), an examination of Berger's list shows that the only candidates are $\frac{SL(17,\mathbb{R})}{SL(16,\mathbb{R})+\mathbb{R}}$, $\frac{SO(17,\mathbb{C})}{SO(16,\mathbb{C})}$ and $\frac{Sp(9,\mathbb{R})}{Sp(1,\mathbb{R})+Sp(8,\mathbb{R})}$ together with

$$\frac{SU^i(n)}{SU^k(k+h)+SU^{i-k}(n-k-h)+T}, \quad (k+h)(n-k-h) = 16,$$

$$\frac{SO^i(n)}{SO^k(k+h)+SO^{i-k}(n-k-h)}, \quad \begin{aligned}(k+h)(n-k-h) &= 32, \\ k+h &> 2, n-k-h > 2,\end{aligned}$$

$$\frac{Sp^i(n)}{Sp^k(k+h)+Sp^{i-k}(n-k-h)+T}, \quad (k+h)(n-k-h) = 8.$$

Now, $\frac{SL(17,\mathbb{R})}{SL(16,\mathbb{R})+\mathbb{R}}$ and $\frac{Sp(9,\mathbb{R})}{Sp(1,\mathbb{R})+Sp(8,\mathbb{R})}$ correspond to the paracomplex and paraquaternionic space forms with the eigenvalue λ of multiplicity 1 and 3, respectively. Proceeding as in the previous case, the other symmetric spaces listed above are either not Osserman or they correspond to the indefinite complex or quaternionic space forms. Therefore, it follows that there exists no semi-Riemannian special Osserman manifold with the eigenvalue λ of multiplicity 15. This completes the proof of Theorem 5.2.1.   □

6. Generalizations and Osserman-Related Conditions

In this chapter we briefly expose some concepts related to the Osserman problem. The exposition will not be so exhaustive as in the previous chapters since our purpose in this chapter is to point out a number of different problems related to the algebraic properties of the curvature tensors of semi-Riemannian manifolds.

As a generalization of the Jacobi operators, we first consider the generalized Jacobi operator associated to a k-plane at each point of a semi-Riemannian manifold. The investigation of the spectral properties of such operators is made in Section 6.1.

An attempt to generalize Osserman condition to affine differential geometry is subject to an additional difficulty that the unit sphere bundle cannot be defined. Therefore, any affine Osserman manifold necessarily has zero eigenvalues for the Jacobi operators. However, such manifolds are not necessarily flat and their properties are investigated in Section 6.2. As an application, other semi-Riemannian Osserman metric tensors are also constructed in the cotangent bundle of a torsion-free affine Osserman manifold (cf. §6.2.3.)

In Section 6.3 we discuss the conjecture of isoparametric geodesic spheres and its relation to the Osserman problem and Lichnerowicz conjecture on harmonic manifolds. A condition on the constancy of the eigenvalues of the Jacobi operators along geodesics is discussed in Section 6.4

Finally, Section 6.5 is devoted to the so-called IP-*spaces*. They are Riemannian manifolds where the skew-symmetric curvature operator $R(x, y)$ has constant eigenvalues.

6.1 Semi-Riemannian Generalized Osserman Manifolds

The notion of generalized Osserman manifold is due to Stanilov and Videv [133], who originally investigated such a condition for the 4-dimensional case. Later on, their results were generalized by Gilkey [67] to arbitrary dimensions and extended to semi-Riemannian geometry in [76].

We start with the Riemannian case. Let $Gr_k(T_pM)$ be the Grassmannian of k-planes in T_pM of a Riemannian manifold (M, g). For each $E \in Gr_k(T_pM)$, let $J(E)$ denote the generalized Jacobi operator

$$J(E) = R(\,\cdot\,,x_1)x_1 + \cdots + R(\,\cdot\,,x_k)x_k,$$

where $\{x_1,\ldots,x_k\}$ is an orthonormal basis for E. (Note that $J(E)$ is independent of the choice of orthonormal basis.)

Definition 6.1.1. *A Riemannian manifold (M,g) is called k-Osserman at $p \in M$ if the characteristic polynomial of $J(E)$ is independent of $E \in Gr_k(T_pM)$, that is, the eigenvalues of $J(E)$ counted with multiplicities are constant for every $E \in Gr_k(T_pM)$. Also, (M,g) is called globally k-Osserman if the characteristic polynomial of $J(E)$ is independent of $E \in \bigcup_{p \in M} Gr_k(T_pM)$.*

Note here that any Riemannian Osserman manifold is 1-Osserman and moreover, real space forms are k-Osserman for all k. However, there exist Riemannian Osserman manifolds which are not k-Osserman. On the other hand, there is a certain kind of duality between the concepts above [67], [133] as pointed out in (2) below.

Theorem 6.1.1. [67] *Let (M,g) be an n-dimensional Riemannian manifold.*

(1) If (M,g) is k-Osserman at $p \in M$ then (M,g) is Einstein at $p \in M$.

(2) If (M,g) is k-Osserman at $p \in M$ then (M,g) is $(n-k)$-Osserman at $p \in M$.

(3) If (M,g) is k-Osserman at $p \in M$ and $2k \neq n$ then (M,g) is 2-stein at $p \in M$.

Proof. (1): Let $\{x_1,x_2\}$ be orthonormal vectors in T_pM and let $\{x_i, i = 1,\ldots,n\}$ be an extension of $\{x_1,x_2\}$ to an orthonormal basis for T_pM. Now let E and $\tilde{E}$ be the k-planes defined by $E = span\{x_1, x_3 \ldots, x_{k+1}\}$ and $\tilde{E} = span\{x_2, x_3 \ldots, x_{k+1}\}$. Since (M,g) is assumed to be k-Osserman at $p \in M$, $J(E) = J(\tilde{E})$, and thus

$$trace\{R_{x_1} + R_{x_3} + \cdots + R_{x_{k+1}}\} = trace\{R_{x_2} + R_{x_3} + \cdots + R_{x_{k+1}}\},$$

and it follows that $Ric(x_1,x_1) = Ric(x_2,x_2)$. Thus (M,g) is Einstein at $p \in M$.

(2): Let E be a k-plane in T_pM and $E^\perp$ be the orthogonal $(n-k)$-plane to E in T_pM. Choose orthonormal bases $\{x_i, i = 1,\ldots,k\}$ and $\{x_i, i = k+1,\ldots,n\}$ for E and $E^\perp$, respectively. Then, since (M,g) is Einstein by (1),

$$J(E) + J(E^\perp) = \sum_{i=1}^{n} R(\,\cdot\,,x_i)x_i = \frac{Sc}{n} Id,$$

where Sc denotes the scalar curvature. Therefore, the eigenvalues of $J(E)$ determine the eigenvalues of $J(E^\perp)$ and hence the claim follows.

(3): To prove that (M,g) is 2-stein at $p \in M$, we need to show that $trace R_{x_1}^{(2)} = trace R_{x_2}^{(2)}$ for any orthonormal vectors $x_1, x_2 \in T_pM$. Let us assume that (M, g) is k-Osserman at $p \in M$ for some $2k \neq n$. Then it follows from (1) and (2) that

$$trace\{J(E)^2\}, \quad trace\{J(E^\perp)^2\} \quad \text{and} \quad trace\{(J(E) + J(E^\perp))^2\}$$

are independent of the k-planes E in T_pM. Thus $trace\{J(E)J(E^\perp)\}$ is also independent of the k-planes E.

Now, let $I = (i_2, \ldots, i_k)$ be a collection of $k - 1$ distinct indices ranging from 2 to n. Also let E_I be the k-plane spanned by $\{x_1, x_{i_2}, \ldots, x_{i_k}\}$. Then

$$\mathcal{E}_{\mathcal{B}}(x_1) = \sum_I trace\{J(E_I)J(E_I^\perp)\}$$

$$= (n - 2)(n - 3) \cdots (n - k) \sum_{i>1} trace\{R_{x_1} R_{x_i}\}$$

$$+ (k - 1)(n - 3) \cdots (n - k) \sum_{\substack{i>1, j>1 \\ i \neq j}} trace\{R_{x_i} R_{x_j}\}$$

$$= (n - 3) \cdots (n - k) \Big\{ \{2(k - 1) - (n - 2)\} trace\{R_{x_1}^{(2)}\} \quad (6.1)$$

$$+ \{(n - 2) - 2(k - 1)\} \sum_{i=1}^{n} trace\{R_{x_1} R_{x_i}\} \quad (6.2)$$

$$- (k - 1) \sum_{i=1}^{n} trace\{R_{x_i}^{(2)}\} \quad (6.3)$$

$$+ (k - 1) \sum_{i=1}^{n} \sum_{j=1}^{n} trace\{R_{x_i} R_{x_j}\} \Big\}. \quad (6.4)$$

Now, since (M, g) is assumed to be k-Osserman at $p \in M$, $\mathcal{E}_{\mathcal{B}}(x_1)$ is independent of $\mathcal{B}$ and x_1. Since (M, g) is Einstein at p, the terms (6.2) and (6.4) are independent of $\mathcal{B}$ and x_1. The term (6.3) depends on $\mathcal{B}$ but not on the distinguished vector x_1. By switching the roles of x_1 and x_2, we obtain $trace R_{x_1}^{(2)} = trace R_{x_2}^{(2)}$ since its coefficient in (6.1) is nonzero. This shows that (M, g) is 2-stein at $p \in M$. $\qquad\square$

Remark 6.1.1. Next, observe that the curvature tensors R^0 and R^J (given in Examples 1.2.1 and 1.2.2) are k-Osserman for all k. Indeed, if $E \in Gr_k(T_pM)$ and $p_r(E) : T_pM \to E$ is the projection, it follows that

$$J_{R^0}(E) = kid - p_r(E) \qquad \text{and} \qquad J_{R^J}(E) = -3p_r(J(E)).$$

Note that, the set of k-Osserman algebraic curvature maps is not a vector space, as pointed out in [76]. In fact, for any k with $2 \leq k \leq n-2$, the algebraic curvature tensor $F = c_0 R^0 + c_1 R^J$ is not k-Osserman unless $c_0 c_1 = 0$.

The above remark shows the existence of Riemannian Osserman manifolds (1-Osserman) which are not k-Osserman for $1 < k < n - 1$. Therefore, it is worth understanding the possible relationships among these notions, as well as to search for a classification of such manifolds which are k-Osserman for all k.

The 2-Osserman condition is studied in [76] and the following is known.

Theorem 6.1.2. *Let (M, g) be a Riemannian globally 2-Osserman manifold. If dim $M = n$ is odd or $n \equiv 2 \mod 4$ then (M, g) is a real space form.*

Proof. First, we show that the 2-Osserman condition at $p \in M$ implies the 1-Osserman condition at $p \in M$.

Let S be the set of eigenvalues of $J(E)$ for E in $Gr_2(T_pM)$. Note that S is a finite set since (M, g) is 2-Osserman at $p \in M$. Let $x \in T_pM$ be a unit vector and choose an orthonormal basis $\{x_1, \ldots, x_{n-1}, x_n\}$ for T_pM with $x_n = x$ such that $R_x(x_i) = \lambda_i x_i$. Now, let $P_i = span\{x_i, x\}$ for all $i < n$. Also since $R_{x_i} x_i = 0$ and $R_x(x_i) = \lambda_i x_i$, it follows that $J(P_i)x_i = \lambda_i x_i$. Thus the eigenvalues λ_i of R_x are in $S \cup \{0\}$ and therefore they are constant. This shows that (M, g) is Osserman at $p \in M$. Hence (M, g) is also globally Osserman, from which it follows by Remark 6.1.1 and Theorems 2.1.4 and 2.1.5 that (M, g) is a real space form.   □

The result of previous theorem has been recently generalized for arbitrary k-Osserman Riemannian manifolds by Gilkey who classify all k-Osserman algebraic curvature tensors for $2 \leq k \leq n - 2$ as follows

Theorem 6.1.3. [70] *Let R be a k-Osserman algebraic curvature tensor for $2 \leq k \leq n - 2$. Then R is a multiple of R^0 provided that n is odd or it is a multiple of R^J for some complex structure if n is even.*

As a consecuence, it follows that the only k-Osserman Riemannian manifolds are the spaces of constant curvature [70].

Motivated by the results in Remark 6.1.1, a weaker generalized Osserman condition can be formulated as follows. An algebraic curvature tensor F on a Hermitian vector space $(V, \langle, \rangle, J)$ is called complex (resp., weakly complex) if $JF(x, y) = F(x, y)J$ (resp., $JF(x, Jx) = F(x, Jx)J$) for all $x, y \in V$. Next, a complex (resp., weakly complex) algebraic curvature tensor is said to be *complex* (resp., *weakly complex*) k-*Osserman* if the eigenvalues of $J_F(E)$ are independent of the k-dimensional complex subspaces $E \subset V$.

Examples of complex and weakly complex k-Osserman algebraic curvature tensors can now be produced as follows:

Theorem 6.1.4. [69] *Suppose there is a Cliff(ν)-module structure on $\mathbb{R}^n$ and consider a set of generators $\{J_1, \ldots, J_\nu\}$ such that $J_iJ_j + J_jJ_i = -2\delta_{ij}$. Use J_1 to give $\mathbb{R}^n$ a complex structure and let $\lambda_0, \ldots, \lambda_\nu$ be arbitrary real numbers. Then,*

(i) $F = \lambda_0 R^0 + \sum_{i=1}^{\nu} \lambda_i R^{J_i}$ is a weakly complex 1-Osserman algebraic curvature tensor on $\mathbb{R}^n$.

(ii) $F = \lambda_0 R^0 + \lambda_1 R^{J_1}$ is a weakly complex k-Osserman algebraic curvature tensor on $\mathbb{R}^n$ for all k.

(iii) $F = R^0 - R^{J_1}$ is a complex k-Osserman algebraic curvature tensor on $\mathbb{R}^n$ for all k.

Note here that the algebraic curvature tensor $F = \lambda_0 R^0 + \lambda_1 R^{J_1}$ is complex if and only if $\lambda_0 + \lambda_1 = 0$ and moreover, $R^0 - R^{J_1}$ has constant holomorphic sectional curvature 1 and, up to a constant, it is the curvature tensor of the Fubini-Study metric tensor on $\mathbb{C}P^{n/2}$. Since the group of complex isometries of $\mathbb{C}P^{n/2}$ acts transitively on the set of complex k-planes in the tangent bundle, it follows that $F = R^0 - R^{J_1}$ is k-Osserman for all k.

The generalized Osserman conditions in Definition 6.1.1 can be extended to semi-Riemannian manifolds as follows. Let (M, g) be a semi-Riemannian manifold with metric tensor of signature (ν, η), and let $Gr_{r,s}(T_pM)$ be the Grassmannian of all subspaces $E \subset T_pM$ such that the restriction of g to E is a nondegenerate inner product of signature (r, s). We assume from now on that $0 \leq r \leq \nu$, $0 \leq s \leq \eta$ and moreover, $(r, s) \neq (0, 0)$ and $(r, s) \neq (\nu, \eta)$. Let $\{x_i\}$ be an orthonormal basis for $E \in Gr_{r,s}(T_pM)$ and define the generalized Jacobi operator by

$$J(E) = \sum_{1 \leq i,j \leq r+s} \varepsilon_{x_i} R_{x_i}.$$

Definition 6.1.2. A semi-Riemannian manifold (M, g) is said to be (r, s)-Osserman at $p \in M$ if the coefficients of the characteristic polynomial of $J(E)$ are independent of $E \in Gr_{r,s}(T_pM)$.

An interesting observation is that only the value $r + s$ is important in the previous definition.

Theorem 6.1.5. Let (M, g) be a semi-Riemannian (r, s)-Osserman manifold at $p \in M$ and let $k = r + s$. Then the coefficients of the characteristic polynomial of $J(E)$ are independent of the nondegenerate $E \in Gr_k(T_pM)$ and (M, g) is $(\bar{r}, \bar{s})$-Osserman for any $(\bar{r}, \bar{s})$ with $\bar{r} + \bar{s} = k$.

Proof. We proceed as in Theorem 1.2.1. Complexify T_pM to $T_p^{\mathbb{C}}M$ and, extend R and g to $T_p^{\mathbb{C}}M$ to be complex linear. Fix k with $1 \leq k \leq n - 1$ and let $\vec{F} = (F_1, \ldots, F_k)$ be a k-vector in $T_p^{\mathbb{C}}M$. Also let $g_{ij}(\vec{F}) = g(F_i, F_j)$. We call $\vec{F}$ a nondegenerate k-frame if $det(g_{ij}(\vec{F})) \neq 0$. Let $\mathcal{F}_k$ be the set of nondegenerate k-frames. Note that since $det(g_{ij}(\vec{F}))$ is a nonconstant polynomial, $\mathcal{F}_k$ is an open connected dense subset of $\mathbb{C}^{nk}$. Now if $\vec{F} \in \mathcal{F} \cap \mathbb{R}^{nk}$, let $E(\vec{F}) = span\{x_1, \ldots, x_k\}$ be the associated nondegenerate real k-plane. We call $\vec{F}$ of type (u, v) if the restriction of g to $E(\vec{F})$ is an inner product of signature (u, v). Let $\mathcal{F}_{uv}$ be the set of real nondegenerate k-frames of type (u, v). For $\overrightarrow{F} \in \mathcal{F}$, let

$$J(\overrightarrow{F})Y = \sum_{1 \leq i,j \leq k} g^{ij} R(Y, x_i)x_j.$$

Then for any $i \geq 1$, the map $c_i : \overrightarrow{F} \mapsto trace\{J(\overrightarrow{F})^i\}$ is an analytic function on $\mathcal{F}$.

Now suppose that (M, g) is (r, s)-Osserman. Then the coefficients of the characteristic polynomial of $J(E(\overrightarrow{F})) = J(\overrightarrow{F})$ are constant on $\mathcal{F}_{rs}$ and hence, $c_i(\overrightarrow{F})$ is constant on $\mathcal{F}_{rs}$. Note that $\mathcal{F} \cap \mathbb{R}^{nk} = \bigcup_{u+v=k} \mathcal{F}_{uv}$. Then since c_i is constant on $\mathcal{F}_{rs}$, c_i is constant on a neighborhood of $\mathcal{F}_{rs}$. Also since $\mathcal{F}$ is connected, c_i is also constant on $\mathcal{F}$ and hence on $\bigcup_{u+v=k} \mathcal{F}_{uv}$. This shows that the coefficients of the characteristic polynomial of the generalized Jacobi operators are constant on $\bigcup_{u+v=k} \mathcal{F}_{uv}$. □

Note here that Theorem 6.1.1 can be generalized to semi-Riemannian manifolds and moreover, one has the following result on the classification of Lorentzian k-Osserman manifolds.

Theorem 6.1.6. *Let (M, g) be a Lorentzian (r, s)-Osserman manifold at each $p \in M$. If $r + s = 2$ or $2r + 2s \neq dim M$ then (M, g) is a real space form.*

Proof. Note that, an analog of Theorem 6.1.1 holds for the semi-Riemannian manifolds (cf. [76]) and thus, if $2r + 2s \neq dim M$, it follows from Theorem 6.1.1-(3) that (M, g) is 2-stein. Now, the result follows since any 2-stein Lorentzian manifold is a real space form.

On the other hand, if $r + s = 2$, proceeding as in Theorem 6.1.2, we show that (M, g) is a Lorentzian Osserman manifold and the result follows from Theorem 3.1.2. □

Following the spirit of Remark 3.1.2, a characterization of k-Osserman Lorentzian manifolds was recently obtained by Stavrov

Theorem 6.1.7. [134] *Let (M, g) be a Lorentzian manifold. Then it is k-Osserman for some $2 \leq k \leq n - 2$ at a point $p \in M$ if and only it the sectional curvature is constant at p.*

Remark 6.1.2. Note here that the possible relations between the 1-Osserman and 2-Osserman conditions are not well understood for indefinite metrics of nonLorentzian signature. It is still not known whether 2-Osserman implies 1-Osserman, since the proof of Theorem 6.1.2 cannot be generalized in a straightforward way.

Moreover, note that some of the Osserman examples studied in §4.1 as well as in §5.1 are k-Osserman for all k. Hence, there is plenty of nonsymmetric examples of semi-Riemannian k-Osserman manifolds with metric tensors of every signature (ν, η) with $\nu, \eta > 1$. However, it has been shown in [26] that any algebraic curvature tensor in dimension 4 satisfying simultaneously the 1 and 2-Osserman conditions must necessarily be of constant curvature or, otherwise, all eigenvalues of the Jacobi operators vanish identically. In the later case, the minimal polynomials of the Jacobi operators must have a double root (no triple roots of the minimal polynomial nor complex eigenvalues of the Jacobi operators may occur.)

6.2 The Osserman Condition in Affine Differential Geometry

In this section, we present some results concerning the study of Osserman problem in affine geometry. The notion of affine Osserman connection originated from the effort to supply new examples of semi-Riemannian Osserman manifolds via the construction called the Riemann extension. This construction assigns to every n-dimensional manifold M with a torsion-free affine connection ∇ a semi-Riemannian metric g_∇ of signature (n, n) on the cotangent bundle T^*M.

6.2.1 Affine Osserman Manifolds

Let M be an n-dimensional manifold with a connection ∇ and let R be the curvature tensor of ∇. We define the *Jacobi operator $R_z : T_pM \to T_pM$ with respect to a vector $z \in T_pM$ by $R_z x = R(x, z)z$.*

Definition 6.2.1. *Let M be an n-dimensional manifold with a connection ∇. Then (M, ∇) is called affine Osserman at $p \in M$ if the characteristic polynomial of R_z is independent of $z \in T_pM$. Also (M, ∇) is called affine Osserman if (M, ∇) is affine Osserman at each $p \in M$.*

Theorem 6.2.1. *Let M be an n-dimensional manifold with a connection ∇. Then, (M, ∇) is affine Osserman at $p \in M$ if and only if the characteristic polynomial of R_z is $p_\lambda(R_z) = \lambda^n$ for every $z \in T_pM$.*

Proof. The "if" part of the proof is obvious. For the proof of the "only if" part, let $z \in T_pM$ and let $p_\lambda(R_z) = \lambda^n + a_{n-1}\lambda^{n-1} + \cdots + a_0$ be the characteristic polynomial of R_z. Then for $c \in \mathbb{R}$, $c \neq 0$, the characteristic polynomial of R_{cz} is

$$p_\lambda(R_{cz}) = det(\lambda I - R_{cz}) = det(\lambda I - c^2 R_z) = c^{2n} det(\tfrac{\lambda}{c^2}I - R_z)$$
$$= \lambda^n + c^2 a_{n-1}\lambda^{n-1} + \cdots + c^{2n}a_0.$$

Hence, since $p_\lambda(R_z) = p_\lambda(R_{cz})$, it follows that $a_{n-1} = \ldots = a_0 = 0$. □

Remark 6.2.1. Note that the concept of affine Osserman manifold is rather restrictive. It is not a generalization of the Riemannian and semi-Riemannian cases because all eigenvalues of the Jacobi operators are shown to be zero (whereas the Jacobi operator itself can still be nontrivial.)

Remark 6.2.2. Note that since $a_{n-1} = -tr R_z = -Ric(z, z)$, it follows from the above theorem that if (M, ∇) is affine Osserman at $p \in M$ then $Ric(z, z) = 0$ for every $z \in T_p M$, that is, the Ricci tensor is skew-symmetric at $p \in M$.

Affine Osserman surfaces are easily characterized by the skew-symmetry of their Ricci tensors as follows:

Theorem 6.2.2. *Let M be a 2-dimensional manifold with a connection ∇. Then, (M, ∇) is affine Osserman at $p \in M$ if and only if the Ricci tensor of ∇ is skew-symmetric at $p \in M$.*

Proof. The "only if" part of the proof is immediate from Remark 6.2.2. To prove the "if" part, note that the characteristic polynomial of the Jacobi operator R_z is of the form $p_\lambda(R_z) = \lambda^2 - Ric(z, z)\lambda + det R_z$. But since Ric is skew-symmetric and $R_z z = 0$, we have that $Ric(z, z) = 0$ and $det R_z = 0$. Hence $p_\lambda(R_z) = \lambda^2$ for every $z \in T_p M$ and it follows from Theorem 6.2.1 that (M, ∇) is affine Osserman at $p \in M$.   □

Affine connections with skew-symmetric Ricci tensors usually present useful anatomy in affine differential geometry (see for example [95] and the references therein.) However, such a condition is quite rigid in dimension two, as shown in the following:

Theorem 6.2.3. *Let M be a 2-dimensional manifold with a torsion-free connection ∇ and assume that (M, ∇) is affine Osserman. If ∇ is locally symmetric, i.e., $\nabla R = 0$, then ∇ is flat, that is, $R = 0$.*

Proof. First note that $\nabla R = 0$ implies that $\nabla Ric = 0$ since $Ric(x, y) = trace\{z \mapsto R(z, x)y\}$. Thus, since Ric is skew-symmetric by Theorem 6.2.2, it follows that the rank of Ric is either zero or two on each connected component of M. But Ric cannot have rank two on any connected component of M, since otherwise Ric is a parallel volume form on this connected component, that is, ∇ is equiaffine on this connected component, in contradiction to the fact that then the Ricci tensor of ∇ should be symmetric. (cf. [111, Prop. 3.1].) Thus $Ric = 0$ on M. But then, since Ric is symmetric, it follows from

$$R(x, y)z = Ric(y, z)x - Ric(x, z)y \qquad (6.5)$$

(cf. [111, p. 18]) so that $R = 0$.   □

Remark 6.2.3. Theorem 6.2.3 fails in higher dimensions. Indeed, consider the connection ∇ on $\mathbb{R}^3$ defined by

$$\nabla_{\frac{\partial}{\partial x_1}}\frac{\partial}{\partial x_1} = \frac{1}{2}\frac{\partial}{\partial x_1}, \quad \nabla_{\frac{\partial}{\partial x_2}}\frac{\partial}{\partial x_2} = -x_2\frac{\partial}{\partial x_2}, \quad \nabla_{\frac{\partial}{\partial x_1}}\frac{\partial}{\partial x_3} = e^{(x_1+\frac{1}{2}x_2^2)}\frac{\partial}{\partial x_2},$$

where (x_1, x_2, x_3) are the usual coordinates on $\mathbb{R}^3$. Then, the only nonvanishing component of the curvature tensor is given by $R(\frac{\partial}{\partial x_1}, \frac{\partial}{\partial x_3})\frac{\partial}{\partial x_1} = e^{(x_1+\frac{1}{2}x_2^2)}\frac{\partial}{\partial x_2}$ and it follows that $(\mathbb{R}^3, \nabla)$ is a nonflat locally symmetric affine Osserman manifold.

Locally homogeneous affine Osserman surfaces were investigated by Kowalski, Opozda and Vlášek, who obtain a complete description of such manifolds as follows.

Theorem 6.2.4. [96] *Let (M, ∇) be a 2-dimensional affine Osserman manifold. If ∇ is torsion-free and locally homogeneous, then one of the following hold:*

A1. *There exists local coordinates (x_1, x_2), $x_1 \neq 0$ such that*

$$\nabla_{\frac{\partial}{\partial x_1}}\frac{\partial}{\partial x_1} = 0$$
$$\nabla_{\frac{\partial}{\partial x_1}}\frac{\partial}{\partial x_2} = -\frac{1}{3}x_1^2\frac{\partial}{\partial x_1} + \frac{1}{x_1}\frac{\partial}{\partial x_2} \tag{6.6}$$
$$\nabla_{\frac{\partial}{\partial x_2}}\frac{\partial}{\partial x_2} = (-\frac{1}{36}x_1^5 + f(x_2)x_2)\frac{\partial}{\partial x_1} - \frac{2}{3}x_1^2\frac{\partial}{\partial x_2}$$

where $f(x_2)$ is an arbitrary function.

A2. *There exists local coordinates (x_1, x_2) such that*

$$\nabla_{\frac{\partial}{\partial x_1}}\frac{\partial}{\partial x_1} = 0$$
$$\nabla_{\frac{\partial}{\partial x_1}}\frac{\partial}{\partial x_2} = x_1\frac{\partial}{\partial x_1} \tag{6.7}$$
$$\nabla_{\frac{\partial}{\partial x_2}}\frac{\partial}{\partial x_2} = (-\frac{1}{2}x_1^2 f_1(x_2) + f_2(x_2))\frac{\partial}{\partial x_1} + (x_1 + f_1(x_2))\frac{\partial}{\partial x_2}$$

where $f_1(x_1)$ and $f_2(x_2)$ are arbitrary functions.

B1. *There exists local coordinates (x_1, x_2), $(x_1 + x_2) \neq 0$ such that*

$$\nabla_{\frac{\partial}{\partial x_1}}\frac{\partial}{\partial x_1} = -\frac{3}{2(x_1+x_2)}\frac{\partial}{\partial x_1} - \frac{1}{2(x_1+x_2)}\frac{\partial}{\partial x_2}$$
$$\nabla_{\frac{\partial}{\partial x_1}}\frac{\partial}{\partial x_2} = -\frac{1}{2k(x_1+x_2)}\frac{\partial}{\partial x_1} - \frac{1}{2(x_1+x_2)}\frac{\partial}{\partial x_2} \tag{6.8}$$
$$\nabla_{\frac{\partial}{\partial x_2}}\frac{\partial}{\partial x_2} = -\frac{1}{2k(x_1+x_2)}\frac{\partial}{\partial x_1} + (1-\frac{1}{2k})\frac{1}{x_1+x_2}\frac{\partial}{\partial x_2}.$$

B2. *There exists local coordinates (x_1, x_2), $-2x_1 + x_2 \neq 0$ such that*

$$\nabla_{\frac{\partial}{\partial x_1}}\frac{\partial}{\partial x_1} = \frac{3}{-2x_1+x_2}\frac{\partial}{\partial x_1} - \frac{2}{-2x_1+x_2}\frac{\partial}{\partial x_2}$$
$$\nabla_{\frac{\partial}{\partial x_1}}\frac{\partial}{\partial x_2} = \frac{1}{-2x_1+x_2}\frac{\partial}{\partial x_2} \tag{6.9}$$
$$\nabla_{\frac{\partial}{\partial x_2}}\frac{\partial}{\partial x_2} = \frac{1}{-2x_1+x_2}\frac{\partial}{\partial x_2}$$

The proof of the above result is based on a detail examination of a PDE system and goes beyond the scope of this monograph. The reader is therefore refer to [96] for a complete proof.

Remark 6.2.4. Affine Osserman connections which are curvature homogeneous have been investigated in [95] where it is shown that, an analytic affine Osserman connection on an analytic two-dimensional manifold is locally homogeneous if and only if it is curvature homogeneous up to order three. Moreover, this bound cannot be improved, as shown in the following: Consider the plane $\mathbb{R}^2$ with usual coordinates (x_1, x_2) and the connection ∇ defined by

$$\nabla_{\frac{\partial}{\partial x_1}} \frac{\partial}{\partial x_1} = 0$$

$$\nabla_{\frac{\partial}{\partial x_1}} \frac{\partial}{\partial x_2} = \nabla_{\frac{\partial}{\partial x_2}} \frac{\partial}{\partial x_1} = e^{x_2} x_1 \frac{\partial}{\partial x_1}$$

$$\nabla_{\frac{\partial}{\partial x_2}} \frac{\partial}{\partial x_2} = \tfrac{1}{2} e^{x_2} x_1^2 \frac{\partial}{\partial x_1} + e^{x_2} x_1 \frac{\partial}{\partial x_2}.$$

Now, a straightforward calculation shows that the Ricci tensor of ∇ is skew-symmetric, and thus, an affine Osserman connection. Moreover, it is shown in [95] that $(\mathbb{R}^2, \nabla)$ admits only a 1-dimensional space of affine Killing vector fields generated by $X = x_1 \frac{\partial}{\partial x_1} - \frac{\partial}{\partial x_2}$, and hence it is not locally homogeneous.

In order to show that $(\mathbb{R}^2, \nabla)$ is curvature homogeneous up to order two, let us choose the origin $(0,0)$ as a fixed point p. Now, after some calculations, it follows that the nonvanishing components of Ric, ∇Ric and $\nabla^2 Ric$ are given by

$$Ric(\tfrac{\partial}{\partial x_1}, \tfrac{\partial}{\partial x_2}) = -e^{x_2},$$

$$(\nabla_{\frac{\partial}{\partial x_2}} Ric)(\tfrac{\partial}{\partial x_1}, \tfrac{\partial}{\partial x_2}) = -1,$$

$$(\nabla^2_{(\frac{\partial}{\partial x_1}, \frac{\partial}{\partial x_2})} Ric)(\tfrac{\partial}{\partial x_1}, \tfrac{\partial}{\partial x_2}) = 2, \qquad (\nabla^2_{(\frac{\partial}{\partial x_2}, \frac{\partial}{\partial x_2})} Ric)(\tfrac{\partial}{\partial x_1}, \tfrac{\partial}{\partial x_2}) = -1.$$

Hence, if x is a point in $\mathbb{R}^2$, then $\Phi_x : T_x\mathbb{R}^2 \to T_p\mathbb{R}^2$ defined by

$$\Phi(\tfrac{\partial}{\partial x_1}) = \tfrac{-e^{x_2}}{2e^{x_2} x_1 - 1} \tfrac{\partial}{\partial x_1} \big|_p$$

$$\Phi(\tfrac{\partial}{\partial x_2}) = \tfrac{x_1 e^{x_2}(3 - 2x_1 e^{x_2})}{2(1 - 2x_1 e^{x_2})} \tfrac{\partial}{\partial x_1} \big|_p + (1 - 2e^{x_2} x_1) \tfrac{\partial}{\partial x_2} \big|_p$$

gives a linear isomorphism between the tangent spaces such that $\Phi^* \nabla^k R = \nabla^k R$ for $k = 0, 1, 2$. Hence, $(\mathbb{R}^2, \nabla)$ is curvature homogeneous up to order two on any subset where $\frac{x_1 e^{x_2}(3 - 2x_1 e^{x_2})}{2(1 - 2x_1 e^{x_2})}$ is well-defined.

6.2.2 Semi-Riemannian Osserman Metric Tensors on Cotangent Bundles

Let M be an n-dimensional manifold and $T_p^* M$ its cotangent space at $p \in M$. The set $T^* M = \bigcup_{p \in M} T_p^* M$ is called the *cotangent bundle* over the manifold

M. Next, we recall some basic concepts on the geometry of the cotangent bundle and refer to [145] for their proofs. A point $\tilde{p} \in T^*M$ is given by $\tilde{p} = (p, \omega)$, where $p \in M$ and $\omega \in T_p^*M$. Also let $\pi : T^*M \to M$, $\pi(\tilde{p}) = p$ be the canonical projection. If f is a function on M, define $f^V = f \circ \pi$ to be its vertical lift to T^*M.

If $(U, (x_i))$ is a chart on M, it induces natural coordinates $(x_i, x_{i'})$ on $\pi^{-1}(U)$, where a 1-form ω on U can be written as $\omega = \sum x_{i'} dx_i$.

For each vector field X on M, define a function $\iota X : T^*M \to \mathbb{R}$ by $\iota X(\tilde{p}) = \iota X(p, \omega) = \omega(X_p)$. In local coordinates, $\iota X(x_i, x_{i'}) = \sum x_{i'} X_i$, where $X = \sum X_i \frac{\partial}{\partial x_i}$. Now, it is important to observe that two vector fields $\tilde{X}, \tilde{Y}$ on T^*M coincide with each other if and only if $\tilde{X}(\iota Z) = \tilde{Y}(\iota Z)$ for all vector fields Z on M. Using this fact, one can define the complete lift of a vector field X on M to a vector field X^C on T^*M by $X^C(\iota Z) = \iota[X, Z]$, for all vector fields Z on M.

It is also important to note that two $(0, s)$-tensor fields $\tilde{T}, \tilde{S}$ on T^*M are the same if and only if $\tilde{T}(X_1^C, \ldots, X_s^C) = \tilde{S}(X_1^C, \ldots, X_s^C)$ for all vector fields $X_1, \ldots, X_s$ on M.

Now, let ∇ be a torsion-free affine connection on M and define its *Riemann extension* to T^*M, which is a metric tensor on T^*M denoted by g_∇, by

$$g_\nabla(X^C, Y^C) = -\iota(\nabla_X Y + \nabla_Y X) \qquad (6.10)$$

for all vector fields X, Y on M.

In the local induced coordinates $(x_i, x_{i'})$ on $\pi^{-1}(U) \subset T^*M$, the Riemann extension is expressed by

$$g_\nabla = \begin{pmatrix} -2x_{k'} \Gamma_{ij}^k & \delta_i^j \\ \delta_i^j & 0 \end{pmatrix}, \qquad (6.11)$$

where Γ_{ij}^k are the Christoffel symbols of the torsion-free connection ∇ with respect to $(U, (x_i))$, which shows that g_∇ is a semi-Riemannian metric tensor of signature (n, n) on T^*M.

Now we have the following:

Theorem 6.2.5. *Let (T^*M, g_∇) be the cotangent bundle of an affine manifold (M, ∇) equipped with the Riemann extension of the torsion-free connection ∇. Then, (T^*M, g_∇) is a semi-Riemannian globally Osserman manifold if and only if (M, ∇) is an affine Osserman manifold.*

Proof. The components of the curvature tensor of (T^*M, g_∇) are related to those of (M, ∇) as follows [145]:

$$\tilde{R}^h_{kji} = R^h_{kji},$$

$$\tilde{R}^{h'}_{kji} = x_{a'}\left\{\nabla_h R^a_{kji} - \nabla_i R^a_{kjh}\right.$$

$$\left. +\Gamma^a_{ht}R^t_{kji} + \Gamma^a_{kt}R^t_{ihj} + \Gamma^a_{jt}R^t_{hik} + \Gamma^a_{it}R^t_{kjh}\right\},$$

$$\tilde{R}^{h'}_{kji'} = -R^i_{kjh}, \qquad \tilde{R}^{h'}_{kj'i} = -R^j_{hik}, \qquad \tilde{R}^{h'}_{k'ji} = -R^k_{hij}.$$

Now, let $\tilde{X} = \alpha_i \frac{\partial}{\partial x_i} + \alpha_{i'} \frac{\partial}{\partial x_{i'}}$ be a vector field on T^*M. Then it follows from the previous formula that the matrix of the Jacobi operator $\tilde{R}_{\tilde{X}}$ with respect to the basis $\{\frac{\partial}{\partial x_i}, \frac{\partial}{\partial x_{i'}}\}$ is of the form

$$\tilde{R}_{\tilde{X}} = \begin{pmatrix} [R_X] & 0 \\ * & {}^t[R_X] \end{pmatrix}, \tag{6.12}$$

where $[R_X]$ is the matrix of the Jacobi operator with respect to the vector field $X = \alpha_i \frac{\partial}{\partial x_i}$ on M.

Now, if (M, ∇) is assumed to be affine Osserman, then R_X has zero eigenvalues for each vector field X on M. Therefore, it follows from (6.12) that the eigenvalues of the Jacobi operators $\tilde{R}_{\tilde{X}}$ vanish for every vector field $\tilde{X}$ on T^*M. Thus (T^*M, g_∇) is globally Osserman.

Conversely, assume that (T^*M, g_∇) is a semi-Riemannian Osserman manifold. If $X = \alpha_i \frac{\partial}{\partial x_i}$ is an arbitrary vector field on M then $\tilde{X} = \alpha_i \frac{\partial}{\partial x_i} + \frac{1}{2\alpha_i}\frac{\partial}{\partial x_{i'}}$ is a unit vector field at every point of the zero section of T^*M. Then from formula (6.12), we see that the characteristic polynomial $p_\lambda(\tilde{R}_{\tilde{X}})$ of $\tilde{R}_{\tilde{X}}$ is the square of the characteristic polynomial $p_\lambda(R_X)$ of R_X. Since for every unit vector field $\tilde{X}$ on T^*M the characteristic polynomial $p_\lambda(\tilde{R}_{\tilde{X}})$ should be the same, it follows that for every vector field X on M the characteristic polynomial $p_\lambda(R_X)$ is the same. Hence (M, ∇) is affine Osserman and, in particular, all eigenvalues are zero. $\square$

In order to apply Theorem 6.2.5 to the construction of semi-Riemannian Osserman metric tensors on contangent bundles, we give a local description of affine Osserman connections in a neighborhood of a point with nonzero curvature tensor.

Theorem 6.2.6. *Let M be a 2-dimensional manifold with torsion-free connection ∇. If (M, ∇) is affine Osserman then at each $p \in M$, either the Ricci tensor of ∇ vanishes or there is a chart (x_1, x_2) in a neighborhood U of p where the nonzero covariant derivatives are given by one of the following three possibilities:*

$$\nabla_{\frac{\partial}{\partial x_1}} \frac{\partial}{\partial x_1} = -\left(\frac{\partial}{\partial x_1}\theta\right)\frac{\partial}{\partial x_1},$$

$$\nabla_{\frac{\partial}{\partial x_2}} \frac{\partial}{\partial x_2} = \left(\frac{\partial}{\partial x_2}\theta\right)\frac{\partial}{\partial x_2},$$

where θ is a smooth function on U such that $\frac{\partial^2}{\partial x_1 \partial x_2}\theta \neq 0$, or

$$\nabla_{\frac{\partial}{\partial x_1}}\frac{\partial}{\partial x_1} = -(\frac{\partial}{\partial x_1}\log\varphi)\frac{\partial}{\partial x_1},$$

$$\nabla_{\frac{\partial}{\partial x_2}}\frac{\partial}{\partial x_2} = \varphi\frac{\partial}{\partial x_1} + (\frac{\partial}{\partial x_2}\log\varphi)\frac{\partial}{\partial x_2},$$

where φ is a smooth function on U such that $\frac{\partial^2}{\partial x_1 \partial x_2}\log\varphi \neq 0$, or

$$\nabla_{\frac{\partial}{\partial x_1}}\frac{\partial}{\partial x_1} = (-\frac{\partial}{\partial x_1}\log\psi + \frac{x_2}{1+x_1 x_2})\frac{\partial}{\partial x_1} + (\frac{1}{\psi(1+x_1 x_2)})\frac{\partial}{\partial x_2}$$

$$\nabla_{\frac{\partial}{\partial x_2}}\frac{\partial}{\partial x_2} = -(\frac{\psi}{1+x_1 x_2})\frac{\partial}{\partial x_1} + (\frac{\partial}{\partial x_2}\log\psi + \frac{x_1}{1+x_1 x_2})\frac{\partial}{\partial x_2},$$

where ψ is a smooth function such that $\frac{\partial^2}{\partial x_1 \partial x_2}\log\psi \neq 0$.

Proof. Note that if the Ricci tensor of ∇ does not vanish at a point $p \in M$ then it defines a volume form in a neighborhood U of p. Therefore there exists a 1-form σ on U such that $\nabla Ric = \sigma \otimes Ric$. This shows that Ric is a recurrent tensor field on U and hence ∇ has recurrent curvature tensor. Now the result follows from [143]. □

Remark 6.2.5. The construction of nonsymmetric Osserman examples in Section 4.1 is based on a family of metric tensors on $\mathbb{R}^4$ defined by (4.1) as follows:

$$g_{(f_1,f_2)} = x_3 f_1(x_1,x_2)dx_1 \otimes dx_1 + x_4 f_2(x_1,x_2)dx_2 \otimes dx_2$$
$$+a[dx_1 \otimes dx_2 + dx_2 \otimes dx_1]$$
$$+b[dx_1 \otimes dx_3 + dx_3 \otimes dx_1 + dx_2 \otimes dx_4 + dx_4 \otimes dx_2]$$

where f_1 and f_2 are smooth functions depending only on the coordinates x_1 and x_2, and satisfying

$$\frac{\partial f_1}{\partial x_2} + \frac{\partial f_2}{\partial x_1} = 0.$$

Now, it follows from (6.11) that, in the case of $a = 0$, $b = 1$, the metric tensors given by (4.1) can be interpreted as those given by the Riemann extensions to $\mathbb{R}^4 = T^*\mathbb{R}^2$ of the torsion-free connection of an affine Osserman manifold corresponding to the case (1) in Theorem 6.2.6. (Note that f_1 and f_2 correspond to $\frac{\partial\theta}{\partial x_1}$ and $\frac{\partial\theta}{\partial x_2}$, respectively, and moreover, the equation above is nothing but the fact that $\frac{\partial^2}{\partial x_2 \partial x_1}\theta = \frac{\partial^2}{\partial x_1 \partial x_2}\theta$).

In Subsection 5.1.2 we constructed a family of nonsymmetric Osserman metric tensors on the tangent bundle of a Riemannian manifold by considering the deformed complete lift metric tensor. It was a specific feature of such examples that all of them are either Type Ia or Type II semi-Riemannian Osserman manifolds when the base manifold M is assumed to be of dimension two. Contrary to that case, Type III Osserman manifolds occur in many cases when considering the Riemann extensions of torsion-free connections of affine Osserman manifolds. In fact, a long but straightforward calculation shows that the Riemann extension of those locally homogenous affine Osserman surfaces discussed in Theorem 6.2.4 are Type III 4-dimensional semi-Riemannian Osserman manifolds.

6.3 Riemannian Manifolds with Isoparametric Geodesic Spheres

A Riemannian manifold is called *harmonic* if the mean curvature of sufficiently small geodesic spheres is a radial function [126]. By generalizing this notion, a Riemannian manifold is called *k-harmonic* if the kth-elementary symmetric functions of the eigenvalues of the shape operators of small geodesic spheres are radial functions. It has been proved by Chen and Vanhecke [38] that 2-harmonicity implies 1-harmonicity (harmonicity.) It is known that compact harmonic manifolds with finite fundamental group are symmetric spaces and hence locally flat or rank-one symmetric spaces [137]. However, it is also known that there exists nonsymmetric harmonic spaces when the assumption on the compactness is dropped [43]. Moreover, it is proven in [141] that Damek-Ricci harmonic spaces are 3-harmonic if and only if they are symmetric (and hence locally rank-one.) This fact, led them to conjecture that two-point homogeneous spaces could be characterized by the property of having isoparametric geodesic spheres.

Clearly any two-point homogeneous space has isoparametric geodesic spheres, since the local isometries act transitively on each sufficiently small geodesic sphere. The converse is not yet known to be true, although some progresses have been made in connection with the Osserman problem. In the rest of this section, we point out that relation.

Let (M, g) be a Riemannian manifold of $\dim M \geq 3$. Let $G_p(r)$ denote the geodesic sphere with radius r centered at the point $p \in M$, and for each point $x \in G_p(r)$, let $B(x, p, r)$ denote the second fundamental form of $G_p(r)$ at x.

Proposition 6.3.1. [78] *Let (M, g) be a Riemannian manifold such that sufficiently small geodesic spheres are isoparametric. Then (M, g) is globally Osserman.*

Proof. Let $p \in M$, $\xi \in S_p(M)$ and $x = exp_p(r\xi)$. Now consider the following series expansion relating the second fundamental form and the Jacobi operator given by

$$B(x, p, r) = \frac{1}{r}Id - \frac{r}{3}R_\xi - \frac{r^2}{4}\nabla_\xi R_\xi + O(r^2),$$

where all the terms on the right are evaluated at the point p. Under the above hypothesis, the eigenvalues of $B(x, p, r)$ are independent of $\xi \in S_p(M)$, and hence

$$h(\xi, p, r) = R_\xi + \frac{3}{4} r \nabla_\xi R_\xi + O(r^2)$$

and any power of the function h is independent of $\xi \in S_p(M)$. In particular,

$$trace(h(\xi, p, r)^k) = trace(R_\xi^k + \frac{3}{4} k r R_\xi^{k-1} \nabla_\xi R_\xi + O(r^2))$$

is independent of $\xi \in S_p(M)$, and hence, $trace R_\xi^k$ and $trace(R_\xi^{k-1} \nabla_\xi R_\xi)$ are independent of $\xi \in S_p(M)$. Thus (M, g) is pointwise Osserman.

Now, defining f_k as in (1.2) we have

$$\xi f_k = \xi trace R_\xi^k$$

is independent of $\xi \in S_p(M)$, and thus, replacing ξ by $-\xi$ shows that $\xi f_k = 0$. Hence the eigenvalues of the Jacobi operators are constant and it follows that (M, g) is globally Osserman. $\qquad \square$

Remark 6.3.1. It follows from the previous theorem that Riemannian manifolds of $\dim M \neq 4k$, $k > 1$ with isoparametric geodesic spheres are either locally flat or rank-one by an application of the results in Chapter 2.

However, beside those cases covered in Chapter 2, no positive (or negative) answer is known about the characterization of Riemannian manifolds with isoparametric geodesic spheres. Further note that, this condition seems to be stronger than the Osserman one, since those manifolds are necessarily harmonic.

Remark 6.3.2. It is possible to make some modifications on conditions under which the conjecture of isoparametric geodesic spheres is stated. (For example, instead of considering geodesic spheres centered at p, one may consider geodesic spheres passing through p.) It is interesting to note that, in any case, the manifold must be Osserman and harmonic. (See [78] for more information.)

6.4 Riemannian $\mathfrak{C}$-manifolds

A further generalization of Osserman condition is given by the so-called $\mathfrak{C}$-spaces. The starting point is the following result.

Theorem 6.4.1. [8] *A Riemannian manifold (M, g) is locally symmetric if and only if for each geodesic γ of (M, g), the following conditions hold:*

($\mathfrak{C}$) *The eigenvalues of the Jacobi operator R_γ are constant along γ,*

($\mathfrak{P}$) *The eigenspaces of the Jacobi operator R_γ are parallel along γ.*

Now, a Riemannian manifold (M, g) is called a $\mathfrak{C}$-*space* if the condition $\mathfrak{C}$ holds for each geodesic of (M, g). Note here that, any Riemannian globally Osserman manifold satisfies condition $\mathfrak{C}$ and moreover, any Riemannian pointwise Osserman manifold satisfying condition $\mathfrak{C}$ is globally Osserman. Yet, there are many nonsymmetric examples among $\mathfrak{C}$-manifolds. For instance, it is shown in [8] that the classes of naturally reductive, commutative and g.o. spaces are $\mathfrak{C}$-manifolds.

If (M, g) is a 2-dimensional Riemannian manifold and γ is a geodesic of (M, g), then the associated Jacobi operator R_γ has two eigenvalue functions along γ which are 0 and $c\kappa$, where c is a constant and κ is the curvature of (M, g) along γ. Hence, a 2-dimensional Riemannian manifold is a $\mathfrak{C}$-space if and only if it is a real space form.

The situation is more complicated in dimension 3, where $\mathfrak{C}$-spaces coincide with the class of D'Atri spaces [94]. They must be locally isometric to one of the following (cf. [8], [92]):

(i) a Riemannian symmetric space,
(ii) $SU(2)$ with a special left-invariant Riemannian metric tensor,
(iii) the universal covering of $SL(2, \mathbb{R})$ with a special left-invariant Riemannian metric tensor,
(iv) the 3-dimensional Heisenberg group with any left-invariant Riemannian metric tensor.

If $\dim M \geq 4$, only some partial results are known. All of them suggest a possible relation between $\mathfrak{C}$-spaces and D'Atri manifolds as follows

Theorem 6.4.2. [35], [130] *Let* (M, g, J) *be a 4-dimensional Kähler manifold. Then the following are equivalent:*

1. (M, g, J) *is a $\mathfrak{C}$-space.*
2. (M, g, J) *is a D'Atri space.*
3. (M, g, J) *is locally symmetric.*

A similar result holds in dimension 5 in the framework of Sasakian geometry [63].

Finally, we note that, contrary to Riemannian pointwise Osserman manifolds, a Riemannian product $(M_1 \times M_2, g_1 \oplus g_2)$ is a $\mathfrak{C}$-space if and only if each factor (M_i, g_i) is a $\mathfrak{C}$-space. (See [8] and [9] for more details and further references.)

6.5 Skew-Symmetric Curvature Operators

Let $\{x, y\}$ be an oriented orthonormal basis for an oriented 2-plane $P = span\{x, y\}$ in the tangent space T_pM of a Riemannian manifold (M, g). Then, the skew-symmetric curvature operator $R_{(P)} = R(x, y)$ is independent of the (oriented) orthonormal basis for P. Motivated by the Osserman problem, this section is devoted to the study of the eigenvalues of $R_{(P)}$ and those manifolds where the eigenvalues of $R_{(P)}$ are independent of P.

Definition 6.5.1. [86] *A Riemannian metric tensor g on a manifold M is called Ivanov-Petrova (or simply IP) at a point $p \in M$ if the eigenvalues of $R_{(P)}$, counted with multiplicities, are constant on the Grassmannian $Gr_2^+(T_pM)$ of oriented 2-planes in T_pM. Also g is called globally IP if the eigenvalues of $R_{(P)}$, counted with multiplicities, are constant on $\bigcup_{p \in M} Gr_2^+(T_pM)$.*

Since any 2-dimensional Riemannian manifold is IP, we start our consideration with the first nontrivial case in dimension 3. Due to the special form of the curvature tensor in dimension 3, a 3-dimensional Riemannian manifold (M, g) is IP (resp., globally IP) if and only if either it is a real space form or two of its principal Ricci curvatures are zero and the third one is a smooth function (resp., constant.) Indeed, if $x, y \in T_pM$, since dim $M = n = 3$, we have

$$\mathrm{trace}R_{(P)}^2 = const. = -2r^2$$

where $P = span\{x, y\}$. Moreover, let $\{x_1, x_2, x_3\}$ be an orthonormal basis for T_pM which consists of eigenvectors of the Ricci operator. Since the curvature tensor is completely determined by the Ricci tensor, one obtains that

$$\kappa^2(x_1, x_2) = \kappa^2(x_1, x_3) = \kappa^2(x_2, x_3) = r^2$$

where $\kappa(x_i, x_j)$ is the curvature of the plane $span\{x_i, x_j\}$. This means that the eigenvalues of the Ricci operator are either $(2\varepsilon a, 2\varepsilon a, 2\varepsilon a)$ or $(0, 0, 2\varepsilon a)$, where $\varepsilon = \pm 1$. Now, if Ω is the subset of M on which the number of distinct eigenvalues of the Ricci operator is locally constant, then the assertion holds.

The local classification of 3-dimensional Riemannian manifolds with principal Ricci curvatures $(0, 0, r(p))$ is still not completely solved. All such Riemannian manifolds are curvature homogeneous but not locally homogeneous and have been described by O. Kowalski in [93].

Example 6.5.1. Motivated by the situation in dimension 3, one has the following examples of Riemannian IP manifolds.

(1) Let (M, g) be a Riemannian real space form. Then the local isometries of (M, g) act transitively on the set of all 2-planes in the tangent bundle of M, and hence (M, g) is globally IP.
    In fact, since the curvature tensor of a real space form is given by $R = cR^0$ as in Example 1.3.1, the eigenvalues of $R(P)$ are $\{\pm\sqrt{-1}c, 0, \ldots, 0\}$, which are independent of P.

(2) Let $M = I \times N$, where I is an open interval of $\mathbb{R}$ and g_N is a metric tensor of constant sectional curvature c on N. Furnish M with the metric tensor $g_M = dt^2 \oplus f(t)g_N$, where the warping function $f(t) = (ct^2 + at + b)/2 > 0$. Then, the eigenvalues of $R(P)$ are $\{\pm\sqrt{-1}C(t), 0, \ldots, 0\}$, where $C(t) =$

$\frac{4cb-a^2}{4f(t)^2}$, which show that (M, g_M) is IP at each point $p \in M$. Moreover, if $a^2 - 4bc \neq 0$, then (M, g_M) is not globally IP.

(3) Let J be a product structure on a vector space V and let $V = V_{(+)} \oplus V_{(-)}$ be the decomposition of V induced by J (see Example 1.2.3.) An inner product $\langle , \rangle$ on (V, J) is said to be "adapted" if the subspaces $V_{(\pm)}$ associated to the eigenvalues ± 1 of J are orthogonal.

Specialize the inner product vector space $(V, J, \langle , \rangle)$ in such a way that the eigenvalue -1 of J has multiplicity 1 and define an algebraic curvature tensor on $(V, J, \langle , \rangle)$ by $R_{J,c}(x, y)z = cR^0(Jx, Jy)z$. Then $R_{J,c}$ is an IP-algebraic curvature tensor .

The approach in investigating the geometrical meaning of IP-condition follows the same spirit as in the Osserman problem. First of all, it focuses on the determination of the possible IP-algebraic curvature tensors and secondly, a local description of IP manifolds is achived after an extensive use of the second Bianchi identity. The results below were firstly obtained by S. Ivanov and I. Petrova in [85] for 4-dimensional manifolds and later extended by P. Gilkey, J. Leahy and H. Sadofsky in [74] to higher dimensions.

In order to describe IP-algebraic curvature tensors, the following notion plays an important role.

Definition 6.5.2. *Let R be an IP-algebraic curvature tensor. Then define the rank of R to be $rank(R) = dim \ Range(R(P))$, where P is an oriented plane.*

The following theorem gives a characterization of IP-algebraic curvature tensors.

Theorem 6.5.1. *[74] If R is an IP-algebraic curvature tensor on $(V^n, \langle , \rangle)$ with $n \geq 5$, $n \neq 7, 8$, then $rank(R) \leq 2$. Moreover an algebraic curvature tensor is IP with $rank(R) = 2$ if and only if $R = R_{J,c}$ as in Example 6.5.1-(3).*

The proof of the above theorem is based on the use of some topological methods. For, let $\mathfrak{so}(\nu)$ denote the Lie algebra of the orthogonal group. Then, $R : Gr_2^+(\mathbb{R}^n) \to \mathfrak{so}(\nu)$ is admissible if R is continuous, $R(-P) = -R(P)$, and dim $ker R(P)$ is constant. If R is admisible then let $rank(R) = dim$ $Range(R(P))$. Now, Theorem 6.5.1 is obtained from the following result.

Lemma 6.5.1. *[74] Let $n \geq 5$ and let $R : Gr_2^+(\mathbb{R}^n) \to \mathfrak{so}(\nu)$ be admissible.*

1. *If $\nu = n$ and if $n \neq 7, 8$, then $rank(R) \leq 2$.*
2. *If $\nu < n$ and if $n \neq 8$ then $R(P) = 0$.*
3. *There exists an admissible $R : Gr_2^+(\mathbb{R}^8) \to \mathfrak{so}(8)$ so that $rank(R) = 8$.*
4. *There exists an admissible $R : Gr_2^+(\mathbb{R}^8) \to \mathfrak{so}(7)$ so that $rank(R) = 6$.*
5. *There exists an admissible $R : Gr_2^+(\mathbb{R}^7) \to \mathfrak{so}(7)$ so that $rank(R) = 6$.*

Remark 6.5.1. Note that in dimension $n = 4$, there is an IP-algebraic curvature tensor which has rank 4. The nonzero components of such algebraic curvature tensor are given by

$$R_{1212} = a_2, \quad R_{1313} = a_2, \quad R_{2424} = a_2,$$
$$R_{1414} = a_1, \quad R_{2323} = a_1, \quad R_{3434} = a_2,$$
$$R_{1234} = a_1, \quad R_{1324} = -a_1, \quad R_{1423} = a_2,$$

where $a_2 + 2a_1 = 0$.

Theorem 6.5.2. *Let (M, g) be an IP manifold.*

1. $\mathrm{rank}(R) = 2$ *everywhere if and only if (M, g) is locally isometric to one of the manifolds in Examples 6.5.1-(1),(2).*
2. *If $\mathrm{rank}(R)$ is at most 2, then either (M, g) is flat or $\mathrm{rank}(R) = 2$.*

Summarizing the previous results, we have the following, which shows that the IP-condition is much more rigid that the Osserman one.

Theorem 6.5.3. [85],[74] *Let (M, g) be a globally IP-manifold. If $\dim M \neq 7$, then (M, g) is either locally isometric to a real space form or a warped product as in Example 6.5.1. Moreover, if $\dim M = n > 4$, $n \neq 7$, then a globally IP-manifold is necessarily a real space form.*

6.5.1 Riemannian Manifolds whose Skew-Symmetric Curvature Operators Have Constant Eigenvalues along each Circle

A smooth curve $c(t)$ on a Riemannian manifold (M, g) parametrized by the arc length with tangent vector field c' is called to be a *circle* of curvature k_1 if its first curvature k_1 is a constant different from zero and all other curvatures are zeros [112]. A *unit circle* is a circle with curvature $k_1 = 1$. For a unit circle,

$$\nabla_{c'} c' = n, \qquad \nabla_{c'} n = -c',$$

where the unit vector field n is the first normal of $c(t)$. All other normals orthogonal to c' and n are parallel along $c(t)$. The differential equations for a unit circle is equivalent to

$$\nabla_{c'} \nabla_{c'} c' + c' = 0.$$

If $\dim M \geq 2$, then for every point $p \in M$ and every two orthonormal vectors $u, v \in T_p M$, there exists locally a unique unit circle $c(t)$ parametrized by the arc length and satisfying the initial conditions;

$$c(0) = p, \qquad c'(0) = u, \qquad (\nabla_{c'} c')(0) = v.$$

Definition 6.5.3. *A Riemannian manifold (M, g) is said to be an $\mathfrak{O}$-space if for any circle $c(t)$, the curvature operator $\kappa_c = R(c', \nabla_{c'} c')$ has constant eigenvalues, counting with multiplicities, along $c(t)$.*

Note that any IP manifold is necessarily a $\mathfrak{O}$-space. However, the classifi-. cation of $\mathfrak{O}$-spaces seems much more difficult and they are classified only in dimension two and three. A 2-dimensional $\mathfrak{O}$-space is necessarily of constant curvature and viceversa. Moreover, in dimension 3, the following result was proved in [84].

Theorem 6.5.4. *Let (M, g) be a 3-dimensional $\mathfrak{O}$-space. Then (M, g) is locally isometric almost everywhere (i.e., on an open and dense subset) to one of the following spaces:*

1. *a real space form*
2. *a Riemannian product of the form $(M \times \mathbb{R}, g \oplus dt^2)$, where (M, g) is a 2-dimensional Riemannian real space form*
3. *a Riemannian manifold with constant principal Ricci curvatures $r_1 = r_2 = 0$, $r_3 \neq 0$.*

Conversely, any 3-dimensional Riemannian manifold as in the above is an $\mathfrak{O}$-space.

References

1. D.V. Alekseevskii, Classification of quaternionic spaces with a transitive solvable group of motions, *Math. USSR-Izv.* **9** (1975), 297–339.
2. D.V. Alekseevskii, B.N. Kimel'fel'd, Structure of homogeneous Riemannian spaces with zero Ricci curvature, *Functional Anal. Appl.* **9** (1975), 97–102.
3. D. M. Alekseevski, N. Blažić, N. Bokan, Z. Rakić, Self-dual and pointwise Osserman spaces, *Arch. Math. (Brno)*, **35** (1999), 193–201..
4. M. Barros, A. Romero, Indefinite Kähler manifolds, *Math. Ann.* **261** (1982), 55–62.
5. J. K. Beem, P. E. Ehrlich, K. L. Easley, *Global Lorentzian Geometry.* (Second Edition) Marcel Dekker, New York, 1996.
6. L. Bérard Bergery, Sur la courbure des métriques riemanniennes invariantes des groupes de Lie et des espaces homogénes, *Ann. Sci. Éc. Norm. Sup.* **11** (1978), 543–576.
7. M. Berger, Les espaces symétriques non compacts, *Ann. Eco. Norm. Sup.* **12** (1957), 85–177.
8. J. Berndt, L. Vanhecke, Two natural generalizations of locally symmetric spaces, *Differential Geom. Appl.* **2** (1992), 57–80.
9. J. Berndt, F. Tricerri, L. Vanhecke, *Generalized Heisenberg groups and Damek-Ricci harmonic spaces*, Lect. Notes in Math. **1598**, Springer-Verlag, Berlin, 1995.
10. A. Besse, *Manifolds all of whose geodesics are closed*, Ergeb. Math. G., Folge **93**, Springer-Verlag, Berlin, Heidelberg, New York, 1978.
11. A. Besse, *Einstein manifolds*, Erg. Math. G., Folge **10**, Springer-Verlag, Berlin, Heidelberg, New York, 1987.
12. R.I. Bishop, S.I. Goldberg, On the topology of positively curved Kähler manifolds, *Tôhoku Math. J.* **15** (1963), 359–364.
13. D. E. Blair, *Contact metric manifolds in Riemannian geometry*, Lect. Notes in Math. **509**, Springer-Verlag, Berlin, 1976.
14. N. Blažić, Paraquaternionic projective space and pseudo-Riemannian geometry, *Publ. Inst. Math. (Beograd)* **60** (74) (1996), 101–107.
15. N. Blažić, N. Bokan, Compact Riemann surfaces with the skew-symmetric Ricci tensor, *Izv. Vyssh. Uchebn. Zaved. Mat.* **9** (1994), 8–12.
16. N. Blažić, N. Bokan, P. Gilkey, A Note on Osserman Lorentzian manifolds, *Bull. London Math. Soc.* **29** (1997), 227–230.
17. N. Blažić, N. Bokan, P. Gilkey, Z. Rakić, Pseudo-Riemannian Osserman manifolds, *Balkan J. Geom. Appl.* **2** (1997), 1–12.
18. N. Blažić, N. Bokan, Z. Rakić, Osserman pseudo-Riemannian manifolds of signature (2, 2), *J. Austr. Math. Soc.*, to appear.
19. N. Blažić, N. Bokan, Z. Rakić, Characterization of type II Osserman manifolds in terms of connection forms, to appear.

20. N. Blažić, N. Bokan, Z. Rakić, Recurrent Osserman spaces, *Bull. Cl. Sci. Math. Nat. Sci. Math.* **23** (1998), 63–69.
21. N. Blažić, N. Bokan, Z. Rakić, The first order PDE system for type III Osserman manifolds, *Publ. de l'Inst. Math. (Beograd)* **62(76)** (1997), 113–119.
22. N. Blažić, N. Bokan, Z. Rakić, Foliation of a dynamically homogeneous neutral manifold, *J. Math. Phys.* **39** (1998), 6118–612.
23. N. Blažić, N. Bokan, Z. Rakić, A note on Osserman conjecture and isotropic covariant derivative of curvature, *Proc. Amer. Math. Soc.* **128** (2000), 245–253.
24. N. Blažić, M. Prvanović, Almost Hermitian Manifolds and Osserman Condition, to appear.
25. E. Boeckx, Einstein-like semi-symmetric spaces, *Arch. Math. (Brno)* **29** (1993), 235–240.
26. A. Bonome, P. Castro, E. García-Río, Generalized Osserman four-dimensional manifolds, *Class. Quantum Grav.* **18** (2001), to appear.
27. A. Bonome, R. Castro, E. García-Río, L. Hervella, On the holomorphic sectional curvature of an indefinite Kähler manifold, *Comp. Rend. Acad. Sci. Paris* **315** (1992), 1183–1187.
28. A. Bonome, R. Castro, E. García-Río, L. Hervella, Curvature of indefinite almost contact manifolds, *J. Geom.* **58** (1997), 66–86.
29. A. Bonome, R. Castro, E. García-Río, L. Hervella, Y. Matsushita, Null holomorphically flat indefinite almost Hermitian manifolds, *Illinois J. Math.* **39** (1995), 635–660..
30. A. Bonome, R. Castro, E. García-Río, L. Hervella, R. Vázquez-Lorenzo, On the paraholomorphic sectional curvature of almost para-Hermitian manifolds, *Houston J. Math.*, **24** (1998), 277–300.
31. A. Bonome, R. Castro, E. García-Río, L. Hervella, R. Vázquez-Lorenzo, Non-symmetric Osserman indefinite Kähler manifolds, *Proc. Amer. Math. Soc.*, **126** (1998), 2763-2769.
32. A. Bonome, R. Castro, E. García-Río, L. Hervella, R. Vázquez-Lorenzo, Pseudo-Riemannian manifolds with simple Jacobi operators, *J. Math. Soc. Japan*, to appear.
33. E. Calabi, Métriques kählériennes et fibrés holomorphes, *Ann. Scient. Ec. Norm. Sup.*, 4e **sér 12** (1979), 269–294.
34. E. Calabi, Isometric Families of Kähler Structures, *The Chern Symposium*, 1979 (W.-Y. Hsiang, S. Kobayashi, I.M. Singer, A. Weinstein, J. Wolf and H.-H. Wu, eds), Springer.Verlag, 1980, pp. 23-39.
35. G. Calvaruso, Ph. Tondeur, L. Vanhecke, Four-dimensional ball-homogeneous and C-spaces, *Beiträge Algebra Geom.* **38** (1997), 325-336.
36. P. Carpenter, A. Gray, T. J. Willmore, The curvature of Einstein symmetric spaces, *Quart. J. Math. Oxford Ser. (2)* **33** (1982), 45–64.
37. J. Cendán-Verdes, E. García-Río, M. E. Vázquez-Abal, On the semi-Riemannian structure of the tangent bundle to a two point homogeneous space, *Riv. Mat. Univ. Parma (5)* **3** (1994), 253–270.
38. B. Y. Chen, L. Vanhecke, Differential geometry of geodesic spheres, *J. Reine Angew. Math.* **325** (1981), 28–67.
39. Q.S. Chi, A curvature characterization of certain locally rank-one symmetric spaces, *J. Diff. Geom.* **28** (1988), 187–202.
40. Q.S. Chi, Quaternionic Kähler manifolds and a curvature characterization of two-point homogeneous spaces, *Illinois J. Math.* **35** (1991), 408–418.
41. Q.S. Chi, Curvature characterization and classification of rank-one symmetric spaces, *Pacific J. Math.* **150** (1991), 31–42.
42. V. Cruceanu, P. Fortuny, P.M. Gadea, A survey on paracomplex geometry, *Rocky Mount. J. Math.* **26** (1996), 83–115.

43. E. Damek, F. Ricci, A class of nonsymmetric harmonic Riemannian spaces, *Bull. Amer. Math. Soc. (N.S.)* **27** (1992), 139–142.

44. M. Dajczer, K. Nomizu, On the boundedness of the Ricci curvature of an indefinite metric, *Bol. Soc. Brasil. Mat.* **11** (1980), 267–272.

45. M. Dajczer, K. Nomizu; On sectional curvature of indefinite metrics II, *Math. Ann.* **247** (1980), 279–282.

46. A. Derdzinski, *Examples de métriques de Kähler et d'Einstein auto-duales sur le plan complexe*, in *Géométrie riemannienne en dimension 4* (Séminaire Arthur Besse, 1978/79), Cedic/Fernand Nathan, Paris, 1981, pp. 334-346.

47. A. Derdzinski, Einstein metrics in dimension four. *Handbook of differential geometry*, **Vol. I**, 419–707, North-Holland, Amsterdam, 2000.

48. I. Dotti, On the curvature of certain extensions of H-type groups, *Proc. Amer. Math. Soc.* **125** (1997), 573–578.

49. I. Dotti, M. J. Druetta, Negatively curved homogeneous Osserman spaces *Differential Geom. Appl.* **11** (1999), 163–178.

50. I. Dotti, M. J. Druetta, Osserman-p spaces of Iwasawa type. *Differential geometry and applications (Brno, 1998)*, 61–72, Masaryk Univ., Brno, 1999.

51. K. L. Duggal, Spacetime manifolds and contact structures, *Int. J. Math. Math. Sci.* **13** (1990), 545–554.

52. B. Fiedler, About the structure of algebraic curvature tensors, to appear.

53. P.M. Gadea, J. Muñoz, Classification of almost para-Hermitian manifolds, *Rend. Mat. Appl.* **11** (1991), 377–396.

54. P.M. Gadea, A. Montesinos, Spaces of constant para-holomorphic sectional curvature, *Pacific J. Math.* **136** (1989), 85–101.

55. K. Galicki, H.B. Lawson, Quaternionic reduction and quaternionic orbifolds, *Maath. Ann.* **282** (1988), 1–21.

56. K. Galicki, T. Nitta, Non-zero scalar curvature generalizations of the ALE hyperkähler metrics, *J. Math. Phys.* **33** (1992), 1765–1771.

57. E. García-Río, D.N. Kupeli, Null and infinitesimal isotropy in semi-Riemannian geometry, *J. Geom. Phys.*, **13** (1994), 207–222.

58. E. García-Río, D.N. Kupeli, 4-Dimensional Osserman Lorentzian manifolds, *New Develop. in Diff. Geom. (Debrecen, 1994)*, 201–211, Math. Appl., **350**, Kluwer Acad. Publ., Dordrecht, 1996.

59. E. García-Río, D.N. Kupeli, M.E. Vázquez-Abal, On a problem of Osserman in Lorentzian geometry, *Differential Geom. Appl.* **7** (1997), 85–100.

60. E. García-Río, D.N. Kupeli, M.E. Vázquez-Abal, R. Vázquez-Lorenzo, Osserman affine connections and their Riemannian extensions, *Differential Geom. Appl.*, **11** (1999), 145–153.

61. E. García-Río, M.E. Vázquez-Abal, R. Vázquez-Lorenzo, Nonsymmetric Osserman pseudo-Riemannian manifolds, *Proc. Amer. Math. Soc.* **126** (1998), 2771–2778.

62. E. García-Río, Y. Matsushita, R. Vázquez-Lorenzo, Paraquaternionic Kähler manifolds, *Rocky Mount. J. Math.* **31** (2001), 237–260.

63. E. García-Río, L. Vanhecke, Five-dimensional ϕ-symmetric spaces, *Balkan J. Geom. Appl.* **1** (1996), 31-44.

64. E. García-Río, R. Vázquez-Lorenzo, Four-dimensional Osserman symmetric spaces, *Geom. Dedicata* **81** (2001), to appear.

65. G.W. Gibbons, C.N. Pope, The positive action conjecture and asymptotically Euclidean metrics in quantum gravity, *Commun. Math. Phys.* **66** (1979), 267–290.

66. P. Gilkey, Manifolds whose curvature operator has constant eigenvalues at the basepoint, *J. Geom. Anal.* **4** (1994), 155–158.

67. P. Gilkey, Generalized Osserman manifolds, *Abh. Math. Sem. Univ. Hamburg* **68** (1998), 125–127.

68. P. Gilkey, Riemannian manifolds whose skew-symmetric curvature operator has constant eigenvalues II, *Differential geometry and applications (Brno, 1998)*, 73–87, Masaryk Univ., Brno, 1999.

69. P. Gilkey, Relating algebraic properties of the curvature tensor to geometry, XII Yugoslav Geometric Seminar (Novi Sad, 1998), *Novi Sad J. Math.* **29** (1999), 109–119.

70. P. Gilkey, Algebraic curvature tensors which are *p*-Osserman, *Differential Geom. Appl.* **14** (2001), 297–311.

71. P. Gilkey, *Geometric Properties of Natural Operators Defined by the Riemannian Curvature Tensor*, to appear

72. P. Gilkey, R. Ivanova, Examples of pseudo-Riemannian complex IP algebraic curvature tensors, to appear.

73. P. Gilkey, R. Ivanova, The Jordan normal form of Osserman algebraic curvature tensors, to appear.

74. P. Gilkey, J.V. Leahy, H. Sadofsky, Riemannian manifolds whose skew-symmetric curvature operator has constant eigenvalues, *Indiana Univ. Math. J.* **48** (1999), 615–634.

75. P. Gilkey, U. Semmelman, Spinors, self-duality, and IP algebraic curvature tensors, to appear.

76. P. Gilkey, G. Stanilov, V. Videv, Pseudo-Riemannian manifolds whose generalized Jacobi operator has constant characteristic polynomial, *J. Geom.* **62** (1998), 144–153.

77. P. Gilkey, I. Stavrov, Curvature tensors whose Jacobi or Szabó operator is nilpotent on null vectors, to appear.

78. P. Gilkey, A. Swann, L. Vanhecke, Isoparametric geodesic spheres and a conjecture of Osserman concerning the Jacobi operator, *Quart. J. Math. Oxford* **46** (1995), 299–320.

79. L. Graves, K. Nomizu, On sectional curvature of indefinite metrics, *Math. Ann.* **232** (1978), 267–272.

80. A. Gray, Classification des variétés approximativement Kählérienness de courbure sectionnelle holomorphe constante, *C. R. Acad. Sci. Paris Sér. A* **279** (1974), 797–800.

81. S. Harris, A Characterization of Robertson-Walker Metrics by Null Sectional Curvature, *Gen. Rel. Grav.* **17** (1985), 493–498.

82. J. Heber, Noncompact homogeneous Einstein spaces, *Invent. math.* **133** (1998), 279–352.

83. S. Ishihara, Quaternion Kählerian manifolds, *J. Diff. Geom.* **9** (1974), 483–500.

84. S. Ivanov, I. Petrova, Riemannian manifolds in which certain curvature operator has constant eigenvalues along each circle, *Ann. Global Anal. Geom.* **15** (1997), 157–171.

85. S. Ivanov, I. Petrova, Riemannian manifold in which the skew-symmetric curvature operator has pointwise constant eigenvalues, *Geom. Dedicata* **70** (1998), 269–282.

86. R. Ivanova, G. Stanilov, A skew-symmetric curvature operator in Riemannian geometry,*Proceedings of the 2nd Gauss Symposium. Conference A: Mathematics and Theoretical Physics (Munich, 1993)*, 391–395, Symposia Gaussiana, de Gruyter, Berlin, 1995.

87. H. Kamada, Y. Machida, Self-duality of metrics of type $(2, 2)$ on four-dimensional manifolds, *Tôhoku Math. J.* **49** (1997), 259–275.

88. H. Kamada, Neutral hyperKähler structures on primary Kodaira surfaces, *Tsukuba J. Math.* **23** (1999), 321–332.

89. H. Karcher, Infinitesimale Charakterisierung von Friedmann-Universen, *Arch. Math. (Basel)* **38** (1982), 58–64.

90. L. Koch-Sen, Infinitesimal null isotropy and Robertson-Walker metrics, *J. Math. Phys.* **26** (1985), 407–410.

91. M. Konishi, On Jacobi fields in quaternion Kähler manifolds with constant Q-sectional curvature, *Hokkaido Math. J.* **4** (1975), 169–178.

92. O. Kowalski, Spaces with volume-preserving symmetries and related classes of Riemannian manifolds, *Rend. Sem. Mat. Politec. Torino*, Fascicolo Speciale (Settembre 1983), 131-158.

93. O. Kowalski, A classification of Riemannian 3-manifolds with constant principal Ricci curvatures $\rho_1 = \rho_2 \neq \rho_3$, *Nagoya Math. J.* **132** (1993), 1-36.

94. O. Kowalski, F. Prüfer, L. Vanhecke, D'Atri spaces, *Topics in geometry*, 241-284, Progr. Nonlinear Differential Equations Appl., **20**, Birkhäuser Boston, Boston, MA, 1996.

95. O. Kowalski, B. Opozda, Z. Vlášek, Curvature homogeneity of affine connections on two-dimensional manifolds *Colloquium Math.* **81** (1999), 123–139.

96. O. Kowalski, B. Opozda, Z. Vlášek, A classification of locally homogeneous affine connections with skew-symmetric Ricci tensor on 2-dimensional manifolds, *Monatsh. Math.* **130** (2000), 109–125.

97. P.B. Kronheimer, The construction of ALE spaces as hyper-Kähler quotients, *J. Diff. Geom.* **29** (1989), 665–683.

98. R.S. Kulkarni, The values of sectional curvature in indefinite metrics, *Comment. Math. Helv.* **54** (1979), 173–176.

99. D.N. Kupeli, On holomorphic and anti-holomorphic sectional curvature of indefinite Kähler manifolds of real dimension $n \geq 6$, *Manuscripta Math.* **80** (1993), 1–12.

100. D.N. Kupeli, On curvatures of indefinite Kähler metrics, *New Zealand J. Math.* **24** (1995), 25–48.

101. D. N. Kupeli, *Singular Semi-Riemannian Geometry*, Kluwer Acad. Publ. Group, Dordrecht, 1996.

102. P. Libermann, Sur le problème d'équivalence de certaines structures infinitésimales, *Ann. Mat. Pura Appl.* **36**, (1954, 27–120).

103. M. A. Magid, Shape operators of Einstein hypersurfaces in indefinite space forms *Proc. Amer. Math. Soc.* **84** (1982), 237–242.

104. S. Marchiavafa, Variétés riemanniennes dont le tenseur de courbure est celuli d'un espace symétrique de rang un, *C. R. Acad. Sci. Paris* **295** (1982), 463–466.

105. Y. Matsushita, Fields of 2-planes on compact simply-connected smooth 4-manifolds *Math. Ann.* **280** (1988), 687–689.

106. Y. Matsushita, Fields of 2-planes and two kinds of almost complex structures on compact 4-dimensional manifolds *Math. Z.* **207** (1991), 281–291

107. Y. Nikolayevsky, Osserman manifolds and Clifford structures, to appear.

108. Y. Nikolayevsky, Two theorems on Osserman manifolds, to appear.

109. K. Nomizu, Conditions for Constancy of the Holomorphic Sectional Curvature, *J. Diff. Geom.* **8** (1973), 335–339.

110. K. Nomizu, Remarks on sectional curvature of an indefinite metric, *Proc. Amer. Math. Soc.* **89** (1983), 473–476.

111. K. Nomizu, T. Sasaki, *Affine Differential Geometry*, Cambridge University Press, Cambridge, 1994.

112. K. Nomizu, K. Yano, On circles and spheres in Riemannian geometry, *Math. Ann.* **210** (1974), 163-170.

113. Z. Olszak, On the existence of generalized complex space forms, *Israel J. Math.* **65** (1989), 214–218.

114. B. O'Neill, *Semi-Riemannian geometry, with applications to relativity*, Academic Press, New York, 1983.
115. V. Oproiu, Harmonic maps between tangent bundles, *Rend. Sem. Mat. Univ. Politecn. Torino* **47** (1989), 47-55.
116. R. Osserman, Curvature in the eighties, *Amer. Math. Monthly* **97** (1990), 731-756.
117. R. Osserman, P. Sarnak, A new curvature invariant and entropy of geodesic flow, *Inventiones Maht.* **77** (1984), 455-462.
118. J.D. Pérez, F.G. Santos, Indefinite quaternion space forms, *Ann. Mat. Pura Appl.* **132** (1982), 383-398.
119. J.D. Pérez, F.G. Santos, F. Urbano, On the axioms of planes in quaternionic geometry, *Ann. Mat. Pura Appl.* **130** (1982), 215-221.
120. J. Petean, Indefinite Kähler-Einstein Metrics on compact complex surfaces , *Commun. Math. Phys.* **189** (1997), 227-235.
121. R. Ponge, H. Reckziegel, Twisted products in pseudo-Riemannian geometry, *Geom. Dedicata* **48** (1993), 15-25.
122. F. Prüfer, F. Tricerri, L. Vanhecke, Curvature invariants, differential operators and local homogeneity, *Trans. Amer. Math. Soc.* **348** (1996) 4643-4652.
123. Z. Rakić, An example of rank two symmetric Osserman space, *Bull. Austral. Math. Soc.* **56** (1997), 517-521.
124. Z. Rakić, On duality principle in Osserman manifolds, *Linear Algebra Appl.* **296** (1999), 183-189.
125. B. A. Rosenfeld, *Geometry of Lie Groups*, Math. Appl. **393**, Kluwer Acad. Publ., Dordrecht, 1997.
126. H. S. Ruse, A. G. Walker, T. J. Willmore, *Harmonic spaces*, Cremonese, Rome, 1961.
127. R.K. Sachs, H. Wu, *General Relativity for Mathematicians*, Springer-Verlag, New York, 1977.
128. S. Salamon, *Riemannian geometry and holonomy groups*, Pitman Research Notes in Math. **201**, New York, 1989.
129. R. Schafer, *An introduction to non-associative algebras*, Pure and Applied Mathematics, Academic Press, New York, 1966.
130. K. Sekigawa, L. Vanhecke, Volume-preserving geodesic symmetries on four-dimensional Kähler manifolds, in *Differential geometry, Peñíscola 1985*, 275-291, Lecture Notes in Math., **1209**, Springer, Berlin-New York, 1986.
131. I. M. Singer, Infinitesimally homogeneous spaces, *Commun. Pure Appl. Math.* **13** (1960), 685-697.
132. G. Stanilov, V. Videv, On Osserman conjecture by characteristic coefficients, *Algebras Groups Geom.* **12** (1995), 157-163.
133. G. Stanilov, V. Videv, Four dimensional pointwise Osserman manifolds, *Abh. Math. Sem. Univ. Hamburg* **68** (1998), 1-6.
134. I. Stavrov, A Note On Generalized Osserman Manifolds, to appear.
135. N. Steenrod, *The topology of fibre bundles*, Princeton, 1965.
136. S. Sternberg, *Lectures on Differential Geometry*, Chelsea Publ. Co., New York, 1983.
137. Z. I. Szabó, The Lichnerowicz conjecture on harmonic manifolds, *J. Differential Geom.* **31** (1990), 1-28.
138. T. Takahashi, Sasakian manifolds with pseudo-Riemannian metric, *Tôhoku Math. J.* **21** (1969), 271-290.
139. J. Thorpe, Curvature and the Petrov Canonical Forms, *J. Math. Phys.* **10** (1969), 1-7.
140. F. Tricerri, L. Vanhecke, Curvature tensors on almost Hermitian manifolds, *Trans. Amer. Math. Soc.* **267** (1981), 365-398.

141. F. Tricerri, L. Vanhecke, Geometry of a class of nonsymmetric harmonic manifolds, *Differential Geometry and Applications (Opava, 1992)*, 415–426, Math. Publ. **1**, Silesian Univ. Opava, Opava 1993.

142. J. Wolf, *Spaces of constant curvature*, Publish or Perish, Boston, Mass., 1974.

143. Y.C. Wong, Two-Dimensional Linear Connexions with Zero Torsion and Recurrent Curvature, *Monatsh. Math.* **68** (1964),175–184.

144. H. Wu, On the de Rham decomposition theorem, *Illinois J. Math.* **8** (1964), 291–311.

145. K. Yano, S. Ishihara, *Tangent and cotangent bundles*, Marcel Dekker, New York, 1973.

146. K. Yano, M. Kon, *Structures on Manifolds*, Series in Pure Math., **3**, World Scientific, Singapore, 1984.

Index

Vol. 1689: W. Fulton, P. Pragacz, Schubert Varieties and Degeneracy Loci. XI, 148 pages. 1998.

Vol. 1690: M. T. Barlow, D. Nualart, Lectures on Probability Theory and Statistics. Editor: P. Bernard. VIII, 237 pages. 1998.

Vol. 1691: R. Bezrukavnikov, M. Finkelberg, V. Schechtman, Factorizable Sheaves and Quantum Groups. X, 282 pages. 1998.

Vol. 1692: T. M. W. Eyre, Quantum Stochastic Calculus and Representations of Lie Superalgebras. IX, 138 pages. 1998.

Vol. 1694: A. Braides, Approximation of Free-Discontinuity Problems. XI, 149 pages. 1998.

Vol. 1695: D. J. Hartfiel, Markov Set-Chains. VIII, 131 pages. 1998.

Vol. 1696: E. Bouscaren (Ed.): Model Theory and Algebraic Geometry. XV, 211 pages. 1998.

Vol. 1697: B. Cockburn, C. Johnson, C.-W. Shu, E. Tadmor, Advanced Numerical Approximation of Nonlinear Hyperbolic Equations. Cetraro, Italy, 1997. Editor: A. Quarteroni. VII, 390 pages. 1998.

Vol. 1698: M. Bhattacharjee, D. Macpherson, R. G. Möller, P. Neumann, Notes on Infinite Permutation Groups. XI, 202 pages. 1998.

Vol. 1699: A. Inoue,Tomita-Takesaki Theory in Algebras of Unbounded Operators. VIII, 241 pages. 1998.

Vol. 1700: W. A. Woyczy´ski, Burgers-KPZ Turbulence, XI, 318 pages. 1998.

Vol. 1701: Ti-Jun Xiao, J. Liang, The Cauchy Problem of Higher Order Abstract Differential Equations, XII, 302 pages. 1998.

Vol. 1702: J. Ma, J. Yong, Forward-Backward Stochastic Differential Equations and Their Applications. XIII, 270 pages. 1999.

Vol. 1703: R. M. Dudley, R. Norvaiša, Differentiability of Six Operators on Nonsmooth Functions and p-Variation. VIII, 272 pages. 1999.

Vol. 1704: H. Tamanoi, Elliptic Genera and Vertex Operator Super-Algebras. VI, 390 pages. 1999.

Vol. 1705: I. Nikolaev, E. Zhuzhoma, Flows in 2-dimensional Manifolds. XIX, 294 pages. 1999.

Vol. 1706: S. Yu. Pilyugin, Shadowing in Dynamical Systems. XVII, 271 pages. 1999.

Vol. 1707: R. Pytlak, Numerical Methods for Optimal Control Problems with State Constraints. XV, 215 pages. 1999.

Vol. 1708: K. Zuo, Representations of Fundamental Groups of Algebraic Varieties. VII, 139 pages. 1999.

Vol. 1709: J. Azéma, M. Émery, M. Ledoux, M. Yor (Eds), Séminaire de Probabilités XXXIII. VIII, 418 pages. 1999.

Vol. 1710: M. Koecher, The Minnesota Notes on Jordan Algebras and Their Applications. IX, 173 pages. 1999.

Vol. 1711: W. Ricker, Operator Algebras Generated by Commuting Projéctions: A Vector Measure Approach. XVII, 159 pages. 1999.

Vol. 1712: N. Schwartz, J. J. Madden, Semi-algebraic Function Rings and Reflectors of Partially Ordered Rings. XI, 279 pages. 1999.

Vol. 1713: F. Bethuel, G. Huisken, S. Müller, K. Steffen, Calculus of Variations and Geometric Evolution Problems. Cetraro, 1996. Editors: S. Hildebrandt, M. Struwe. VII, 293 pages. 1999.

Vol. 1714: O. Diekmann, R. Durrett, K. P. Hadeler, P. K. Maini, H. L. Smith, Mathematics Inspired by Biology. Martina Franca, 1997. Editors: V. Capasso, O. Diekmann. VII, 268 pages. 1999.

Vol. 1715: N. V. Krylov, M. Röckner, J. Zabczyk, Stochastic PDE's and Kolmogorov Equations in Infinite Dimensions. Cetraro, 1998. Editor: G. Da Prato. VIII, 239 pages. 1999.

Vol. 1716: J. Coates, R. Greenberg, K. A. Ribet, K. Rubin, Arithmetic Theory of Elliptic Curves. Cetraro, 1997. Editor: C. Viola. VIII, 260 pages. 1999.

Vol. 1717: J. Bertoin, F. Martinelli, Y. Peres, Lectures on Probability Theory and Statistics. Saint-Flour, 1997. Editor: P. Bernard. IX, 291 pages. 1999.

Vol. 1718: A. Eberle, Uniqueness and Non-Uniqueness of Semigroups Generated by Singular Diffusion Operators. VIII, 262 pages. 1999.

Vol. 1719: K. R. Meyer, Periodic Solutions of the N-Body Problem. IX, 144 pages. 1999.

Vol. 1720: D. Elworthy, Y. Le Jan, X-M. Li, On the Geometry of Diffusion Operators and Stochastic Flows. IV, 118 pages. 1999.

Vol. 1721: A. Iarrobino, V. Kanev, Power Sums, Gorenstein Algebras, and Determinantal Loci. XXVII, 345 pages. 1999.

Vol. 1722: R. McCutcheon, Elemental Methods in Ergodic Ramsey Theory. VI, 160 pages. 1999.

Vol. 1723: J. P. Croisille, C. Lebeau, Diffraction by an Immersed Elastic Wedge. VI, 134 pages. 1999.

Vol. 1724: V. N. Kolokoltsov, Semiclassical Analysis for Diffusions and Stochastic Processes. VIII, 347 pages. 2000.

Vol. 1725: D. A. Wolf-Gladrow, Lattice-Gas Cellular Automata and Lattice Boltzmann Models. IX, 308 pages. 2000.

Vol. 1726: V. Marić, Regular Variation and Differential Equations. X, 127 pages. 2000.

Vol. 1727: P. Kravanja M. Van Barel, Computing the Zeros of Analytic Functions. VII, 111 pages. 2000.

Vol. 1728: K. Gatermann Computer Algebra Methods for Equivariant Dynamical Systems. XV, 153 pages. 2000.

Vol. 1729: J. Azéma, M. Émery, M. Ledoux, M. Yor Séminaire de Probabilités XXXIV. VI, 431 pages. 2000.

Vol. 1730: S. Graf, H. Luschgy, Foundations of Quantization for Probability Distributions. X, 230 pages. 2000.

Vol. 1731: T. Hsu, Quilts: Central Extensions, Braid Actions, and Finite Groups. XII, 185 pages. 2000.

Vol. 1732: K. Keller, Invariant Factors, Julia Equivalences and the (Abstract) Mandelbrot Set. X, 206 pages. 2000.

Vol. 1733: K. Ritter, Average-Case Analysis of Numerical Problems. IX, 254 pages. 2000.

Vol. 1734: M. Espedal, A. Fasano, A. Mikelić, Filtration in Porous Media and Industrial Applications. Cetraro 1998. Editor: A. Fasano. 2000.

Vol. 1735: D. Yafaev, Scattering Theory: Some Old and New Problems. XVI, 169 pages. 2000.

Vol. 1736: B. O. Turesson, Nonlinear Potential Theory and Weighted Sobolev Spaces. XIV, 173 pages. 2000.

Vol. 1737: S. Wakabayashi, Classical Microlocal Analysis in the Space of Hyperfunctions. VIII, 367 pages. 2000.

Vol. 1738: M. Émery, A. Nemirovski, D. Voiculescu, Lectures on Probability Theory and Statistics. XI, 356 pages. 2000.

Vol. 1739: R. Burkard, P. Deuflhard, A. Jameson, J.-L. Lions, G. Strang, Computational Mathematics Driven by Industrial Problems. Martina Franca, 1999. Editors: V. Capasso, H. Engl, J. Periaux. VII, 418 pages. 2000.

Recent Reprints and New Editions

4. Lecture Notes are printed by photo-offset from the master-copy delivered in camera-ready form by the authors. Springer-Verlag provides technical instructions for the preparation of manuscripts. Macro packages in T_EX, L^AT_EX2e, $L^AT_EX2.09$ are available from Springer's web-pages at

http://www.springer.de/math/authors/b-tex.html.

Careful preparation of the manuscripts will help keep production time short and ensure satisfactory appearance of the finished book.

The actual production of a Lecture Notes volume takes approximately 12 weeks.

5. Authors receive a total of 50 free copies of their volume, but no royalties. They are entitled to a discount of 33.3 % on the price of Springer books purchase for their personal use, if ordering directly from Springer-Verlag.

Commitment to publish is made by letter of intent rather than by signing a formal contract. Springer-Verlag secures the copyright for each volume. Authors are free to reuse material contained in their LNM volumes in later publications: A brief written (or e-mail) request for formal permission is sufficient.

Addresses:

Professor J.-M. Morel
CMLA, Ecole Normale Supérieure de Cachan
61 Avenue du Président Wilson
94235 Cachan Cedex France
E-mail: Jean-Michel.Morel@cmla.ens-cachan.fr

Professor B. Teissier
Université Paris 7
UFR de Mathématiques
Equipe Géométrie et Dynamique
Case 7012
2 place Jussieu
75251 Paris Cedex 05
E-mail: Teissier@ens.fr

Professor F. Takens, Mathematisch Instituut,
Rijksuniversiteit Groningen, Postbus 800,
9700 AV Groningen, The Netherlands
E-mail: F.Takens@math.rug.nl

Springer-Verlag, Mathematics Editorial, Tiergartenstr. 17
D-69121 Heidelberg, Germany
Tel.: *49 (6221) 487-701
Fax: *49 (6221) 487-355
E-mail: lnm@Springer.de